ROCK FORMATIONS AND UNUSUAL GEOLOGIC STRUCTURES

THE LIVING EARTH

ROCK FORMATIONS AND UNUSUAL GEOLOGIC STRUCTURES

EXPLORING THE EARTH'S SURFACE

REVISED EDITION

JON ERICKSON

FOREWORD BY ERNEST H. MULLER, PH.D.

Checkmark Books®
An imprint of Facts On File, Inc.

ROCK FORMATIONS AND UNUSUAL GEOLOGIC STRUCTURES
Exploring the Earth's Surface, Revised Edition

For information contact:

Checkmark Books
An imprint of Facts On File, Inc.
11 Penn Plaza
New York, NY 10001

Library of Congress Cataloging-in-Publication Data

Erickson, Jon, 1948–
Rock formations and unusual geologic structures : exploring the earth's surface / by Jon Erickson—Rev. ed.
p. cm. — (The living earth)
Includes bibliographical references and index.
ISBN 0-8160-4328-0 (hardcover: alk. paper) ISBN 0-8160-4590-9 (pbk)
1. Geology. 2. Geomorphology. I. Title.

QE33 .E75 2001
550—dc21 00-049038

Checkmark Books are available at special discounts when purchased in bulk quantities for businesses, associations, institutions or sales promotions. Please contact our Special Sales Department at 212/967-8800 or 800/322-8755.

You can find Facts On File on the World Wide Web at http://www.factsonfile.com

Text design by Cathy Rincon
Cover design by Nora Wertz
Illustrations by Dale Williams, Jeremy Eagle, and Dale Dyer, © Facts On File

Printed in the United States of America

MP Hermitage 10 9 8 7 6 5 4 3 2 1
(pbk) 10 9 8 7 6 5 4 3 2 1

This book is printed on acid-free paper.

CONTENTS

TABLES

ACKNOWLEDGMENTS

The author thanks the National Aeronautics and Space Administration (NASA), the National Optical Astronomy Observatories (NOAO), the U.S. Army, the U.S. Army Corps of Engineers, the U.S. Department of Agriculture-Forest Service, the U.S. Department of Agriculture-Soil Conservation Service, the U.S. Department of Energy, the U.S. Geological Survey (USGS), and the U.S. Navy for providing photographs for this book.

The author also thanks Frank K. Darmstadt, Senior Editor, and Cynthia Yazbek, Associate Editor, for their assistance in the preparation of this book.

FOREWORD

ROCK FORMATIONS AND UNUSUAL GEOLOGIC STRUCTURES

Today people travel farther, faster, and more often than in past generations. This might well afford superb opportunity for making better acquaintance with wonders of the world around us. However, because we are in a hurry we often blindly miss out on the wealth of pleasure and information to be gained from rocks and landscapes.

At time scales far greater than limits in everyday experience, our Earth is dynamic. Its components are constantly undergoing change. Weathering of rocks is recognizable between observations. Valleys are occasionally eroded overnight. Mountain ridges are elevated and worn down at rates so gradual as to escape ordinary perception. Through geologic time, ocean basins have changed in depth and configuration and slow drift has altered the shapes and relative positions of continents. Constant transformation in geologic time is unmistakable, and the evidence is at hand. Even inconspicuous rock fragments bear evidence of the past, available for compilation and correlation by geologists, but also to be noted and appreciated by alert and interested nonprofessionals.

It is to the latter audience that Erickson's *Rock Formations and Unusual Geologic Structures* is directed as an introduction to geology, to alert the untutored but zealous observer to awareness of the origins, nature, and relevance of clues inherent in rocks and rock structures. This revised edition begins with grand-scale generalizations on differentiation of Earth's crust and dynamics of plate tectonism that account for continental masses and ocean basins. Ensuing chapters visit complex landscapes on truncated ancient igneous and metamorphic terranes. More gratifying for detailed analysis are sedimentary regions, products of weathering and erosion of older rocks followed by deposition in beds or strata. Even when complicated by folding and faulting, correlation of such beds in separate exposures can be facilitated by fossil content and reference to a previously described type section. Finally treated are unique features such as impact craters, rift lines, erosion remnants, caverns, and collapsed structures.

—Ernest H. Muller, Ph.D.

INTRODUCTION

Among the most impressive features our planet has to offer are its abounding rock formations and geologic structures. The Earth's surface is covered by a thin veneer of sediments that provides an impressive terrain of ragged mountains and jagged canyons. The most fascinating geologic features were created by erosion. No other terrain can compare with mountain ranges, created by the forces of uplift and erosion. Some of the most magnificent scenery was carved from solid rock by running water and flowing ice. The removal of sediment by wind erosion scours the land, producing deflation basins and blowouts.

The most spectacular examples of the handiwork of groundwater are caves. Over geologic time, water has dissolved great quantities of soluble rock, forming extensive mazes of tunnels in the Earth's crust. Calderas form when the roof of a magma chamber collapses or when a powerful volcano decapitates itself, resulting in a broad depression. The dissolution of subsurface materials or the withdrawal of underground fluids causes the surface to sink, while other ground failures occur when subterranean sediments liquefy during earthquakes or violent volcanic eruptions.

The surface of the Earth is sculpted by a number of forces, providing a cornucopia of unusual geological structures, ranging from narrow spires, mesas, ragged crags, and tall pillars carved out of stone; dikes and volcanic necks created by ancient volcanoes; arches and caves eroded out of solid rock; and meteorite and volcanic craters. The Earth possesses a variety of holes in the ground, including potholes, sinkholes, and numerous craters. Among the most unique depressions are kimberlite pipes, fumaroles and geysers, crater lakes, and lava lakes. These are just a few of the many wonders the Earth's active geology has to offer.

The text chronicles the formation of the Earth's crustal rocks and their interactions with each other. It then examines the weathering, erosional, and sedimentary processes that create landforms. It discusses the type sections that define the Earth's geologic features, including dating, correlating, and mapping geologic formations. It explains the forces that shape the Earth's crust, including the folding and faulting of the Earth's rocks. It examines the different types of igneous activity that shape the planet, including volcanoes and other igneous processes.

After an examination of the basic building blocks of the Earth's crust, the following chapters focus attention on specific types of rock formations and geologic structures. Some major depressions sculptured into the Earth include canyons, rift zones, trenches, valleys, and basins. The geologic features that define the arid and coastal regions, the landforms resulting from glacial erosion and deposition, and caves along with related structures are examined in great detail. Ground failures and collapsed structures and their impact on the Earth's surface is discussed, as well as craters formed by meteorite impacts. The final chapter visits unique rock formations and geologic structures formed by unusual geologic activities.

This revised and updated edition is a much expanded examination of rock formations and geologic structures. Science enthusiasts will particularly enjoy this fascinating subject and gain a better understanding of how the forces of nature operate on Earth. Students of geology and earth science will also find this a valuable reference book to further their studies. Readers will enjoy this clear and easily readable text that is well illustrated with evocative photographs, detailed illustrations, and helpful tables. A comprehensive glossary is provided to define difficult terms, and a bibliography lists references for further reading. The geologic processes that shape the surface of our planet are an example of the spectacular forces that create the living Earth.

1

THE EARTH'S CRUST

THE FORMATION OF CONTINENTS

This chapter examines the formation of crustal rocks, including the shields, cratons, and terranes that comprise the continents. The Earth, as with all the terrestrial planets, Mercury, Venus, and Mars, has a central core, surrounded by an intermediate layer, or mantle, and covered by a thin shell called the crust. The study of metallic meteorites, which once formed the cores of early planetoids that have disintegrated, suggests the Earth's core is composed of iron and nickel. The age of the Earth, estimated at 4.6 billion years, is based on an agreement between the ages of meteorites thought to have formed at the same time as the planet. Furthermore, the ages of lunar rocks generally agree with the Earth's oldest rocks, which formed about 4 billion years ago when the crust first segregated from the mantle.

In the Earth's early stages of formation, it was ceaselessly pounded by numerous giant meteorites. During this time, the planet was struck by as many as three Mars-sized bodies. One of these impactors might have been responsible for creating the Moon by launching great quantities of material into Earth orbit, where it coalesced into a satellite, the largest in the solar system relative to its mother planet. The giant impacts also might have initiated the formation of continents by melting massive quantities of basalt. The continents grew quite rapidly, forming in a burst of creation starting as early as 4.2 billion years ago. The Earth is unique among planets because it is the only one known to have distinct continents, which are essentially thick slabs of granite riding on a sea of semimolten rock in the upper mantle.

Underlying all other rocks on the Earth's surface is a thick layer of basement complex, composed of ancient granitic and metamorphic rocks that have been in existence for nine-tenths of Earth history. These rocks form the nuclei of the continents and first appeared during a period of mantle segregation and outgassing, which created the crust along with the atmosphere and ocean. One remarkable feature about these rocks is that despite their great age they are similar to more recent rocks, signifying that geologic processes began quite early and had a long and productive life.

PRECAMBRIAN SHIELDS

At the beginning of the Archean eon from about 4 billion to 2.5 billion years ago, the interior rocks that make up the Earth's mantle gradually began to cool. This resulted in the creation of a permanent crust composed of a thin layer of basalt lava flows that erupted on the surface long before the ocean basins began to fill with water. Embedded in the thin basaltic crust were granitic blocks that assembled into microcontinents. These were lighter than the basalt, which enabled them to remain on the surface, drifting freely in the convection currents of the mantle as it pushed or pulled them along.

Slices of granitic crust combined into stable bodies of basement rock, upon which all other rocks were deposited. The basement rocks formed the nuclei of the continents and are presently exposed in broad, low-lying domelike structures called shields. The shields are extensive uplifted areas essentially bare of recent sedimentary deposits. Many shields such as the Canadian Shield (Fig. 1), which covers most of eastern Canada and extends down into Wisconsin and Minnesota, are fully exposed in areas that were ground down by flowing ice sheets during the ice ages. The exposure of the Canadian shield from Manitoba to Ontario is attributed to uplifting of the crust by a mantle plume and the erosion of sediments in the uplifted area. Some of the oldest known rocks of North America are the 2.5-billion-year-old granites of the Canadian Shield.

Many of the best exposures of ancient rocks in the United States are the 1.8-billion-year-old metamorphic rocks of the Vishnu Schist on the bottom of the Grand Canyon (Fig. 2). In northern Arizona, over a mile of sedimentary rocks overlie the bedrock of the Grand Canyon. The oldest of these rocks is about 800 million years old, leaving a billion years of geologic history unac-

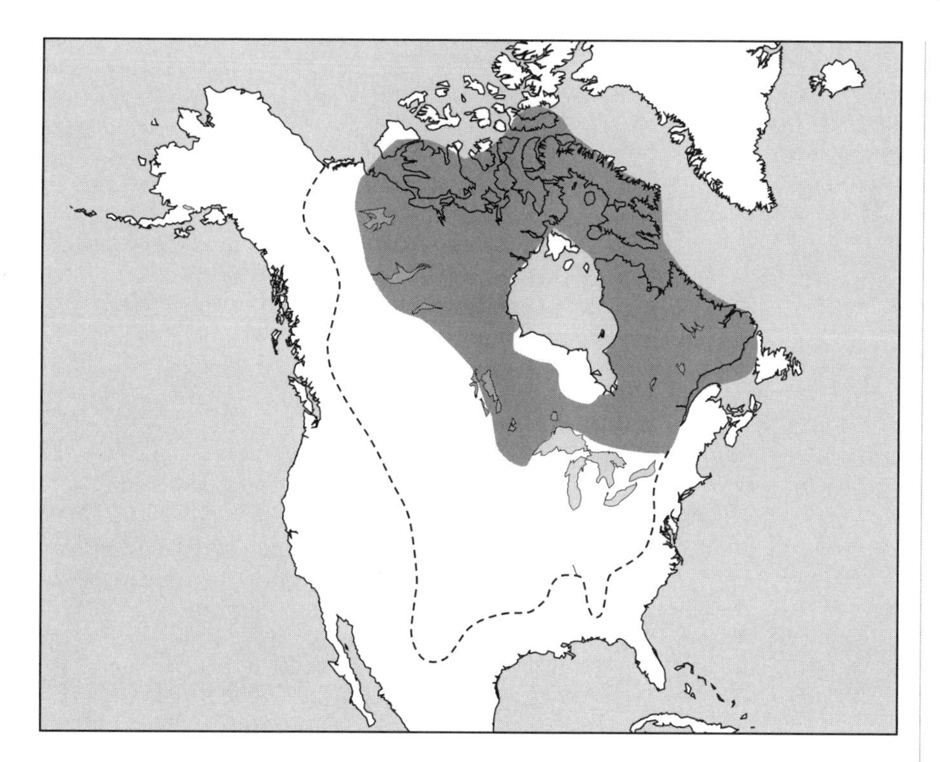

Figure 1 The Canadian Shield (darkened area) and platforms (enclosed in dashed line).

counted. During this time, the floor of the Grand Canyon was worn down by erosion, creating a gap in time known as a hiatus.

Archean greenstone belts (Fig. 3) are dispersed among and around the shields. They consist of a jumble of metamorphosed (recrystallized) lava flows and sediments possibly derived from island arcs (chains of volcanic islands on the edges of subduction zones) caught between colliding continents. Their green color is derived from chlorite, a greenish micalike mineral. The existence of greenstone belts is evidence that plate tectonics might have operated as early as the Archean.

Ophiolites, from the Greek *ophis* meaning "serpent," due to their mottled green color, date as far back as 3.6 billion years. These rocks were also caught in the greenstone belts, as slices of ocean floor were shoved up on the continents by drifting plates. Therefore, ophiolites are among the best evidence for ancient plate motions. They are vertical cross sections of oceanic crust peeled off during plate collisions and plastered onto the continents. This produced a linear formation of greenish volcanic rocks along with light-colored masses of granite and gneiss, common igneous and metamorphic rocks. Pillow lavas, tubular bodies of basalt extruded undersea, are also found in the greenstone belts, signifying these volcanic eruptions took place on the ocean floor. Many ophiolites contain ore-bearing rocks that are important mineral resources throughout the world.

Figure 2 *The Grand Canyon, showing Precambrian Vishnu Schist overlain by the younger rocks of the Grand Canyon Series and Tonto Group, Coconino County, Arizona.*

(Photo by E. D. McKee, courtesy USGS)

Greenstone belts are found in all parts of the world and occupy the ancient cores of the continents. They span an area of several hundred square miles and are surrounded by an immense expanse of gneiss, the metamorphic equivalent of granite rocks and the predominant Archean rock type. The best-known greenstone belt is perhaps the Swaziland sequence in the Barberton Mountain Land of southeastern Africa. It is more than 3 billion years old and reaches nearly 12 miles thick.

Greenstone belts are of particular interest to geologists, not only as evidence for Archean plate tectonics, but also because they contain most of the world's gold deposits. Most South African gold mines are in greenstone belts, and the Kolar belt in India holds the richest gold deposits in the world. The area lies in southern India, which comprises a 3-billion-year-old craton, one of the first pieces of continental crust to appear. Since greenstone belts are essentially Archean in age, their disappearance from the geologic record around 2.5 billion years ago marks the end of the Archean eon.

The abundance of chert in deposits older than 2.5 billion years indicates that most of the crust was deeply submerged during this time. Cherts are dense, extremely hard sedimentary rocks composed of microscopic grains of silica. Most Precambrian cherts are thought to be chemical sediments precipitated from silica-rich water in deep oceans. Ancient cherts dating 3.5 billion years old also contain microfilaments, believed to have a bacterial origin and therefore are among the oldest occurrences of life.

The 3.8-billion-year-old metamorphosed marine sediments of the Isua Formation in a remote mountainous area in southwest Greenland (Fig. 4) pro-

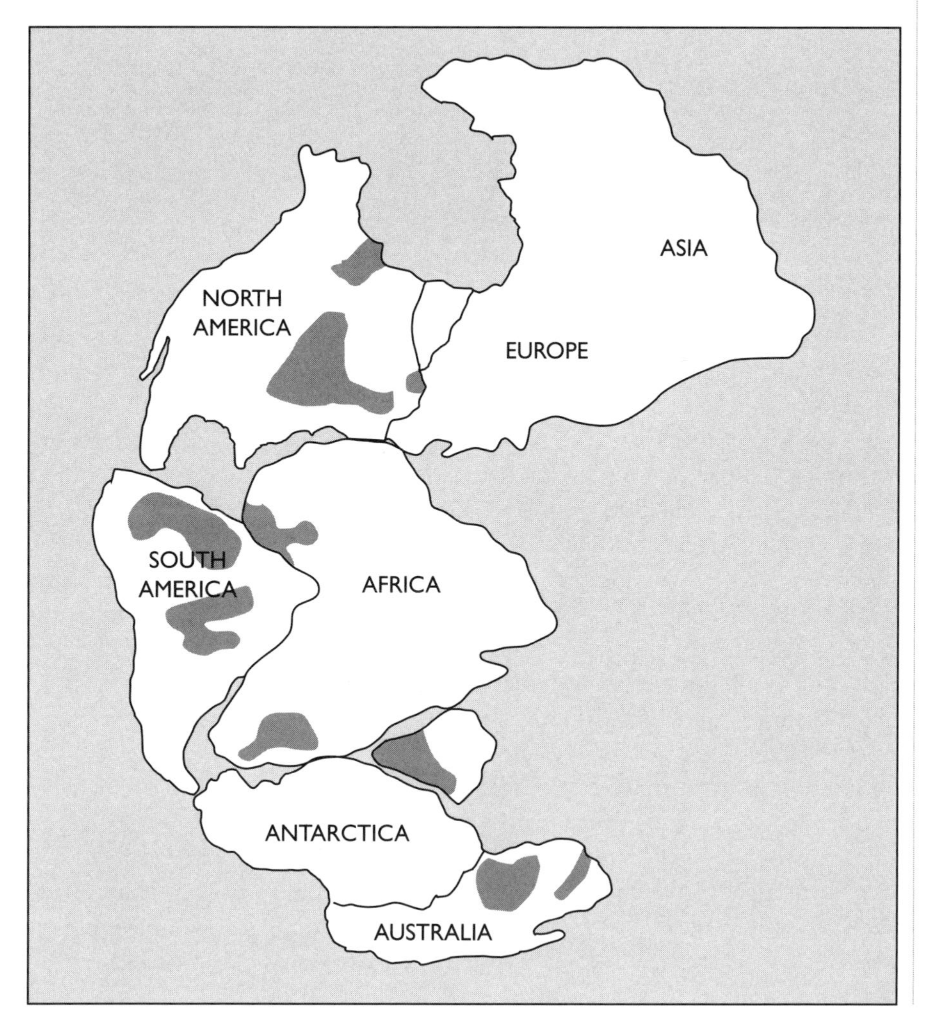

Figure 3 *Archean greenstone belts comprise the ancient cores of the continents.*

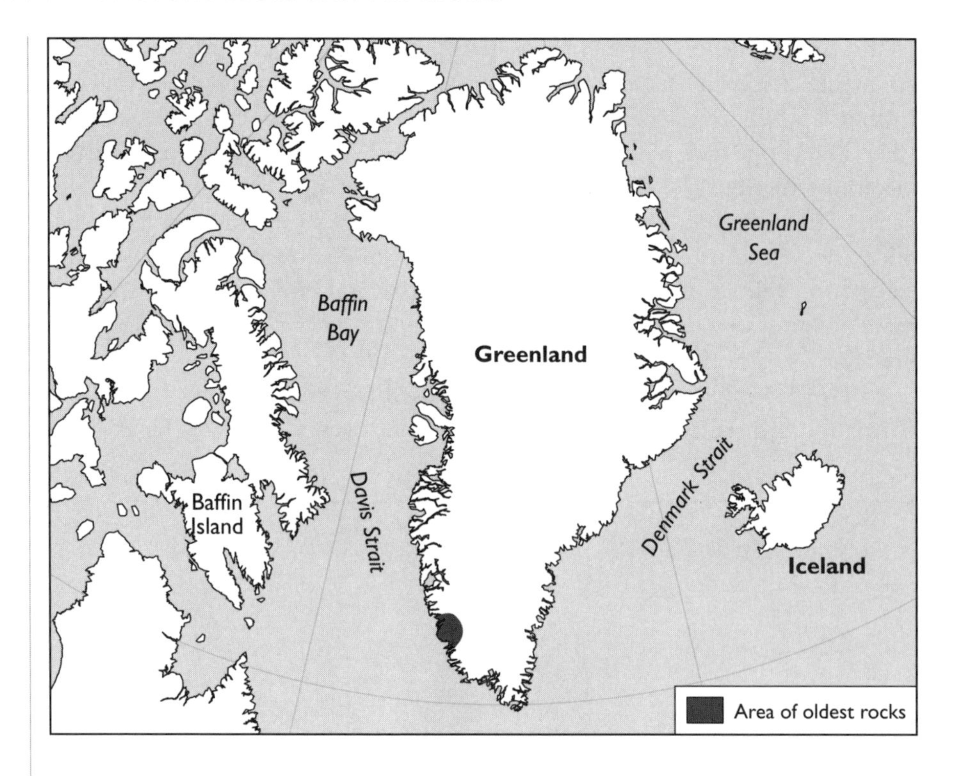

Figure 4 *Location of the Isua Formation in southwestern Greenland, which contains some of the oldest rocks on Earth.*

vide evidence for oceans during this time. The seas contained abundant dissolved silica, which leached out of volcanic rock pouring unto the ocean floor. Modern seawater is deficient in silica because organisms such as sponges and diatoms extract it to build their skeletons. When the organisms die, their skeletons build massive deposits of diatomaceous earth.

The Proterozoic eon from 2.5 billion to 570 million years ago marked a major change from the turbulent Archean. At the beginning of the Proterozoic, much of the present landmass was in existence. The continental crust had grown to an average thickness of between 15 and 25 miles, near what it is today. Most material presently locked up in sedimentary rocks was already at or near the surface, with ample sources of Archean rocks for erosion and redeposition.

Most Proterozoic sediments comprise sandstones and siltstones derived from Archean greenstones. Conglomerates, consolidated equivalents of sand and gravel, were particularly abundant in the Proterozoic. The Proterozoic is also known for its widespread terrestrial redbeds, so-named because their sediment grains were cemented by red iron oxide. Their appearance around 1 billion years ago signifies the atmosphere and ocean contained significant quantities of oxygen by this time. Between about 1 billion years and 550 million years ago, the oxygen content of the atmosphere increased from less than 2 percent to 20 percent.

The weathering of primary rocks also yielded solutions of calcium carbonate, magnesium carbonate, calcium sulfate, and sodium chloride, which precipitated consecutively into limestone, dolomite, gypsum, and halite. The Mackenzie Mountains of northwest Canada contain dolomite deposits 6,500 feet thick. These were mainly formed by chemical precipitates rather than caused by biologic activity because shell-producing organisms had not yet evolved. Carbonate rocks such as limestone and chalk, formed chiefly by organic processes involving shells and skeletons of simple organisms, became much more common during the late Proterozoic, beginning about 700 million years ago. Earlier, these rocks were relatively rare due to the scarcity of lime-secreting organisms.

CRATONS

The shields are surrounded by continental platforms, which are broad, shallow depressions of basement rock covered by nearly flat-lying sedimentary rocks. The shields and platforms together form the cratons (Fig. 5), which were the first pieces of land to appear and are found in the interiors of all continents. They are composed of ancient igneous and metamorphic rocks remarkably similar in composition to those of today, indicating that a rock cycle was fully in place by the beginning of the Proterozoic.

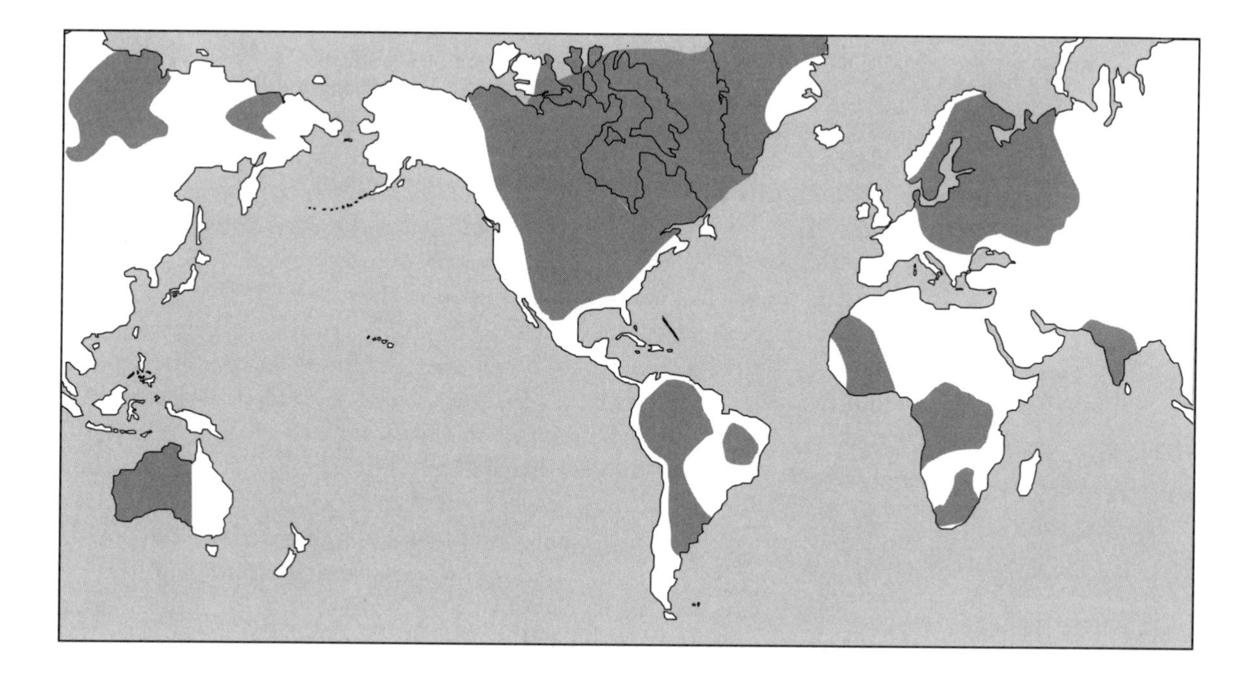

Figure 5 *The worldwide distribution of stable cratons that make up the continents.*

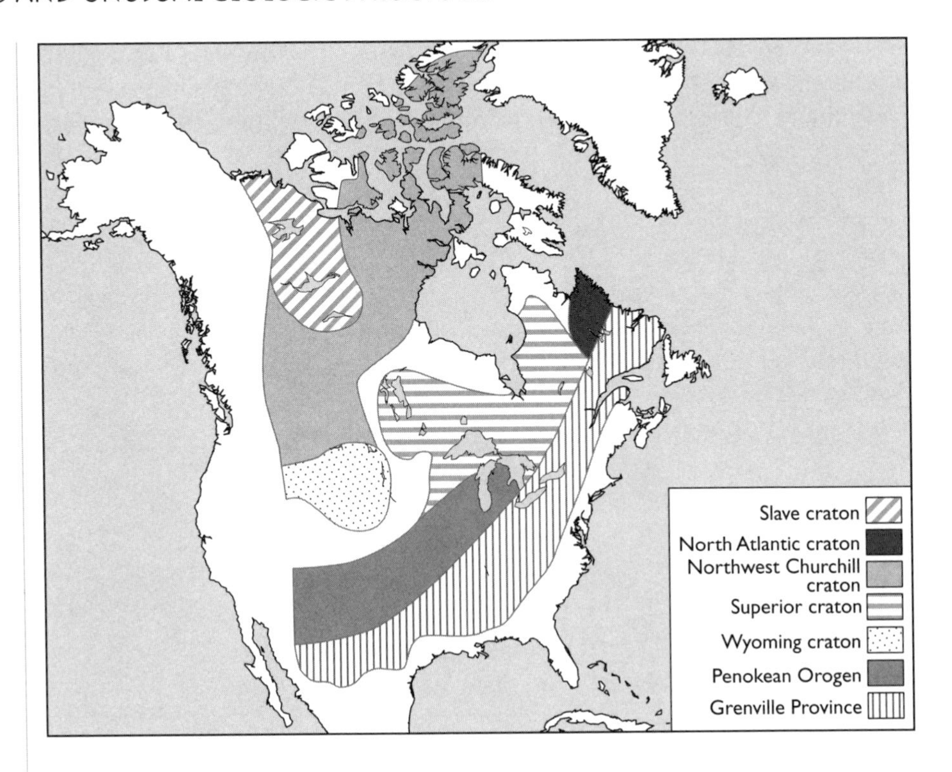

Figure 6 *The cratons that make up the North American continent came together some 2 billion years ago.*

The cratons were slivers of crust that collided with and bounced off each other. As the Earth aged and continued to cool, the cratons slowed their erratic wanderings and began to stick together, forming over a dozen protocontinents. They are more than 2.5 billion years old and range in size from smaller than the state of Texas to about half the area of the United States. As a whole, however, they comprised only about a tenth of today's total landmass.

Continents were more stable and combined into a supercontinent composed of Archean cratons, granitic crust forming the cores of the continents. Many Archean cratons throughout the world were assembled around the same time, with the original cratons forming within the first 1.5 billion years of the Earth's existence. The North American continent consists of seven cratons, comprising central Canada and north-central United States (Fig. 6), that assembled around 2 billion years ago, making it the oldest continent. The African and South American continents did not aggregate until about 700 million years ago. Over the past half-billion years, about a dozen individual continental plates welded together to form Eurasia (Fig. 7). It is the youngest and largest modern continent and is still being pieced together with chunks of crust arriving from the south, riding on highly mobile tectonic plates.

At Cape Smith on the Hudson Bay lies a 2-billion-year-old slice of oceanic crust squeezed onto the land, a telltale sign that continents collided and closed an ancient ocean. Arcs of volcanic rock also weave through central and eastern Canada down into the Dakotas. In a region between Canada's Great Bear Lake and the Beaufort Sea lies the roots of an ancient mountain range that run through the basement rock. The mountains were formed by the collision of North America and another landmass between 1.2 and 0.9 billion years ago.

Continental collisions continued to add a large area of new crust to the growing proto-North American continent. The better part of the continental crust underlying the United States from Arizona to the Great Lakes to Alabama formed in one great surge of crustal generation between 1.9 and 1.7 billion years ago, unequaled in North America since. The assembled North American continent was stable enough to resist a billion years of jostling and rifting and continued to grow as bits and pieces of continents and island arcs were plastered to its margins. An example is the Superior province just north of the Great Lakes, which consists of island arcs and sediment sandwiched together and broken around the edges.

This rapid growth marked the most energetic period of tectonic activity and crustal generation in Earth history. The presence of volcanic rock near the eastern edge of North America is a sign that a giant rift ripped through

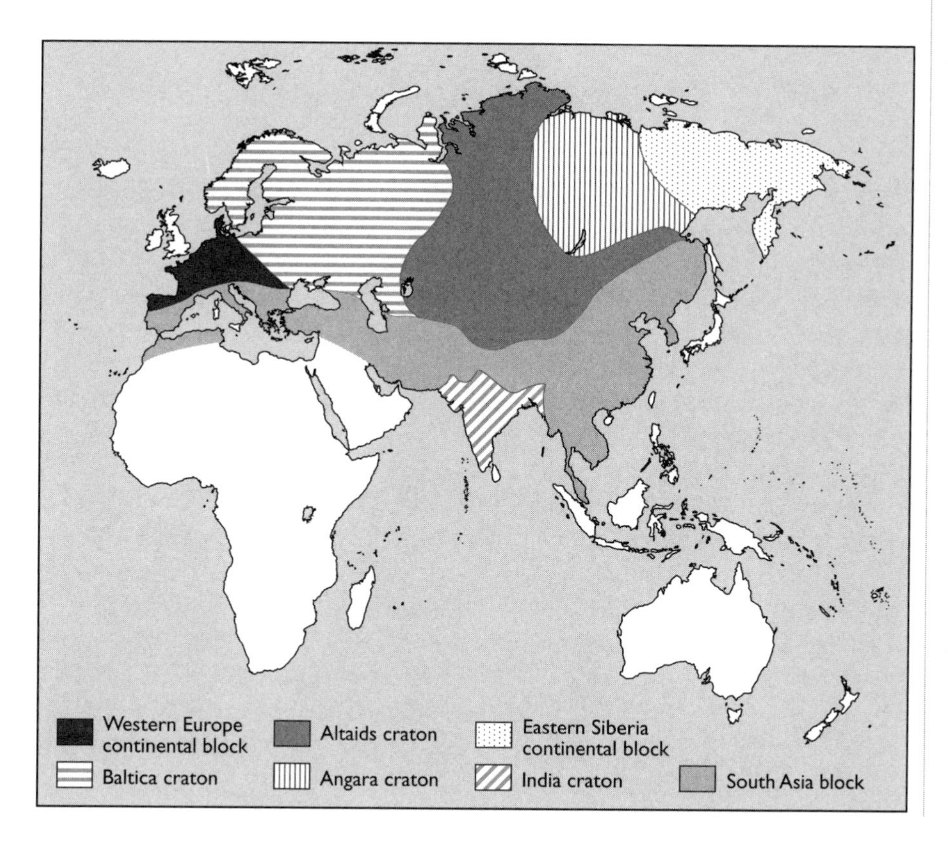

Figure 7 *The major cratons of Eurasia.*

the continent when it was part of a supercontinent in the early Proterozoic. Toward the end of the Proterozoic around 600 million years ago, another supercontinent named Rodinia, Russian for "motherland," with North America at its core broke apart into possibly four or five major continents, although their geographies were different from what they are today. As a prelude to this continental breakup, about 1.1 billion years ago, a great volcano-filled rift valley split the continent from what is now Kansas to Lake Superior.

The cores of the world's mountain ranges also contain very ancient rocks, which were once buried deep in the crust but have since been raised to the surface (Fig. 8). As the mountains were pushed upward, huge blocks of granite were thrust up by tectonic forces originating deep within the Earth. When the continents collided, they crumpled the crust, raising mountain

Figure 8 *The Rocky Mountains represent upraised portions of the Earth's interior.*

(Photo by George A. Grant, courtesy National Park Service)

Figure 9 *An outcrop of retrograde blueschist rocks in the Seward Peninsula region, Alaska.*

(Photo by C. L. Sainsbury, courtesy USGS)

ranges at the point of contact. The sutures joining the landmasses are still visible as cores of ancient mountains more than 2 billion years old.

Caught between the converging cratons was an assortment of debris swept up by the drifting continents, including sediments from continental shelves and the ocean floor, stringers of volcanic rock, and small scraps of continents, fractured by faults. In addition, pieces of ocean crust called ophiolites were thrust upon the land along with blueschists (Fig. 9), which are metamorphosed rocks of subducted ocean crust shoved up on the continents.

The cratons contain the world's oldest rocks and date as far back as 4 billion years. They are composed of highly altered granite and metamorphosed marine sediments and lava flows. The rocks originated from intrusions of magma into the primitive ocean crust. The magma cooled slowly and separated into a light component, which rose toward the surface, and a heavy component, which settled to the bottom of the magma chamber. Some magma also seeped through the crust, where it poured out as lava on the ocean floor. Successive intrusions and extrusions of magma built up the crust until it finally broke the surface of a global sea.

The cratons were highly mobile and moved about freely on the semimolten rocks of the asthenosphere, the fluid portion of the upper mantle. The independent minicontinents periodically collided with each other. The collisions crumpled the leading edges of the cratons, forming small parallel mountain ranges perhaps only a few hundred feet high. Volcanoes were highly active on the cratons as well, and lava and ash continued to build the landmasses upward and outward. New crustal material was also added to the interior of the cratons by magmatic intrusions composed of molten crustal rocks recycled through the upper mantle. This cooled the mantle, slowing the cratons. As the cratons grew more sluggish, they developed a greater tendency to cling to each other. All the cratons eventually coalesced into a single large landmass several thousand miles wide.

TERRANES

The cratons are patchwork mixtures, consisting of crustal pieces known as terranes (Fig. 10), which were assembled into geologic collages. The term *terrane* should not be confused with *terrain,* which means "landform". Terranes are usually bounded by faults and are distinct from their geologic surroundings. The boundary between two or more terranes is called a suture zone. The composition of terranes generally resembles that of an oceanic island or plateau. Others are composed of a consolidated conglomerate of pebbles, sand, and silt that accumulated in an ocean basin between colliding crustal fragments.

Terranes exist in a variety of shapes and sizes from small slivers to subcontinents such as India, which is a single great terrane. Most terranes are elongated bodies that deform when they collide and attach to a continent. The assemblage of terranes in China is being stretched and displaced in an east-west direction due to the continuing squeeze India is exerting on southern Asia as it raises the Himalayas. Granulite terranes are high-temperature meta-

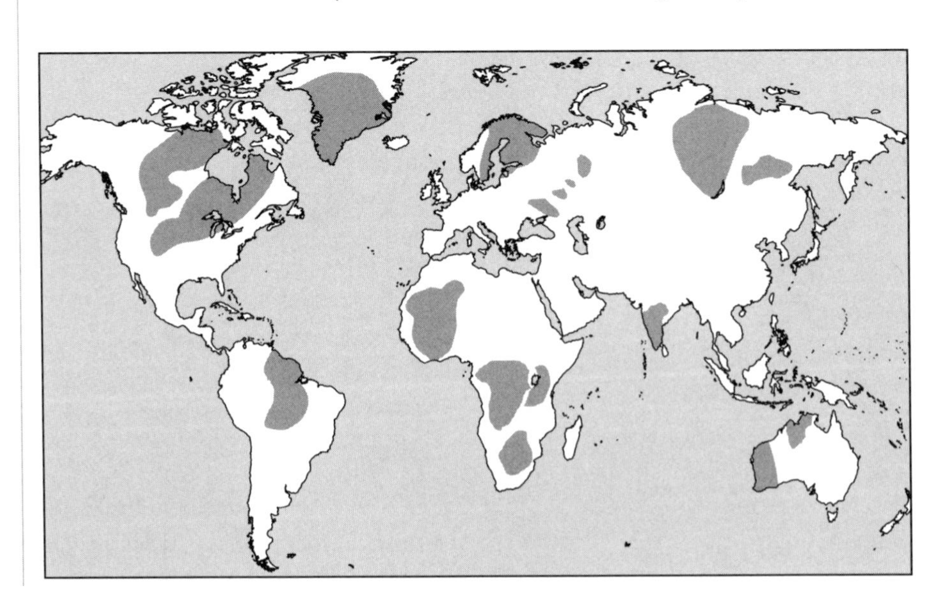

Figure 10 *The distribution of 2-billion-year-old terranes.*

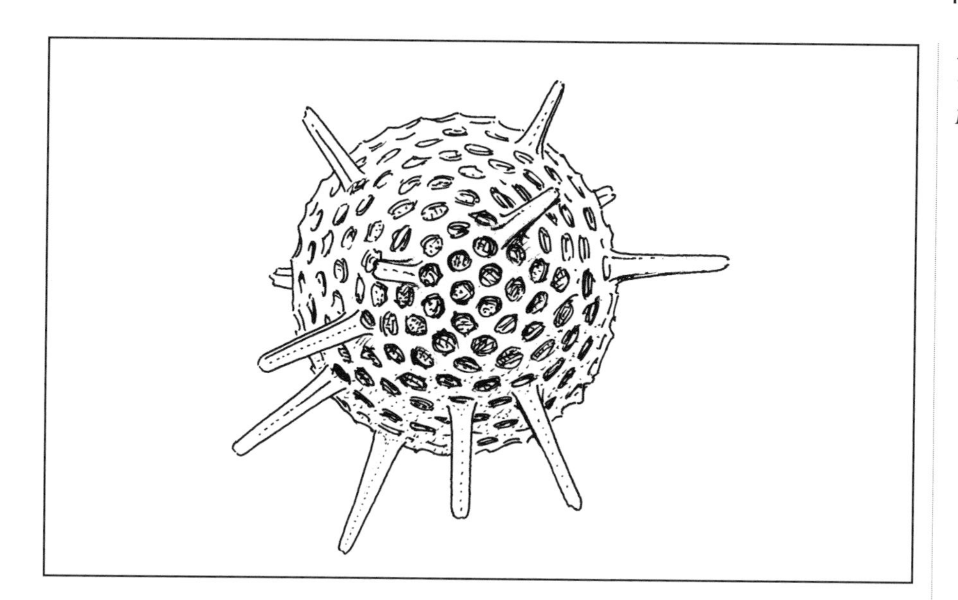

Figure 11 *Radiolarians were marine planktonic protozoans.*

morphic belts, which formed in the deeper parts of continental rifts. They also comprise the roots of mountain belts formed by continental collision, such as the Alps and Himalayas. North of the Himalayas is a belt of ophiolites, which marks the boundary between the sutured continents. Terrane boundaries are commonly marked by ophiolite belts, consisting of marine sedimentary rocks, pillow basalts, sheeted-dike complexes, gabbros, and peridotites.

The terranes range in age from less than 200 million to more than a billion years old. Their ages are determined by the study of entrained fossil radiolarians (Fig. 11), marine protozoans that built skeletons of silica and were abundant from about 500 million to 160 million years ago. Different species also defined specific regions of the ocean where the terranes originated.

Suspect terranes, so-named because of their exotic origins, are fault-bounded blocks with geologic histories apart from those of neighboring terranes and of adjoining continental masses. Suspect terranes were displaced over great distance before finally being accreted to a continental margin. Some North American suspect terranes have a western Pacific origin and were displaced thousands of miles to the east.

Many terranes in western North America have rotated clockwise as much as 70 degrees or more, with the oldest terranes having the greatest rotations. Terranes created on an oceanic plate retain their shapes until they collide and accrete to a continent. They are then subjected to crustal movements that modify their overall dimensions.

Until about 250 million years ago, the western edge of North America ended near present-day Salt Lake City. Over the last 200 million years, North America has expanded by more than 25 percent during a major pulse of

Figure 12 *Steeply dipping Paleozoic rocks of the Brooks Range, near the head of the Itkillik River east of Anaktuvuk Pass, Northern Alaska.*

(Photo by J. C. Reed, courtesy USGS)

crustal growth. Much of western North America was assembled from oceanic island arcs and other crustal debris skimmed off the Pacific plate as the North American plate headed westward. Northern California is a jumble of crust assembled some 200 million years ago. A nearly complete slice of ocean crust at least 2.7 billion years old shoved up on the continents by drifting plates sits in the middle of Wyoming.

The entire state of Alaska is an agglomeration of terranes comprising pieces of an ancient ocean that preceded the Pacific called the Panthalassa. The terranes that make up the Brooks Range, the spine of northern Alaska, are great sheets stacked one on top of another (Fig. 12). The entire state is an assemblage of some 50 terranes set adrift over the past 160 million years by the wanderings and collisions of crustal plates, parts of which are still arriving from the south. California west of the San Andreas Fault has been drifting northward for millions of years, and in another 50 million years it will wander as far north as Alaska, adding another piece to the puzzle.

A large portion of the Alaskan panhandle, known as the Alexander terrane, began as part of eastern Australia some 500 million years ago. About 375 million years ago, it broke free from Australia, traversed the Pacific Ocean, stopped briefly at the coast of Peru, sliced past California swiping part of the Mother Lode gold belt, and crashed into the North American continent around 100 million years ago.

The actual distances terranes can travel varies considerably. Basaltic seamounts that accreted to the margin of Oregon moved from nearby offshore, while similar rock formations around San Francisco, California, came from halfway across the Pacific Ocean. The city itself is built on three different and distinct rock units. At their usual rate of travel, terranes could make a complete circuit of the globe in only about half a billion years.

The accreted terranes also played a major role in the creation of mountain chains along convergent continental margins. For example, the Andes appeared to have been raised by the accretion of oceanic plateaus along the continental margin of South America. Along the mountain ranges in western North America the terranes are elongated bodies due to the slicing of the crust by a network of northwest-trending faults. One of these is the San Andreas Fault in California, which has undergone some 200 miles of displacement in the last 25 million years.

CRYSTALLINE ROCK

The first rocks to form on Earth were igneous, from Latin meaning "fire." Forming directly from molten magma, most igneous rocks arise from new material in the mantle, some are derived from the subduction of oceanic crust into the mantle, and others originate from the melting of continental crust. The first two types continuously build the continents, while the latter type adds nothing to the total volume of continental crust.

Igneous rocks are mostly silicates, which are compounds of silica and oxygen that contain metal ions. They are not simple chemical compounds, however, because their composition is not determined by a fixed ratio of atoms. Often two or more compounds are present in a solid solution with each other. In this manner, the components can be mixed in any ratio over a wide range.

If magma extrudes onto the Earth's surface either by a fissure eruption, the most prevalent kind, or a volcanic eruption, which builds majestic mountains (Fig. 13), it produces a variety of rock types depending on the source material, which in turn controls the type of eruption. Ejecta from volcanoes has a wide range of chemical, mineral, and physical properties. Nearly all volcanic products are silicate rocks, containing various amounts of other elements. Basalts are relatively low in silica with a high content of calcium, magnesium, and iron. Magmas with larger amounts of silica, sodium, and potassium along

Figure 13 *Mount Hood Volcano, Cascade Range, Hood River County, Oregon.*

(Photo by M.V. Shulters, courtesy of USGS)

with lesser amounts of magnesium and iron form rhyolites, with mostly quartz grains, and andesites, with mostly feldspar grains.

Igneous rocks are classified by their mineral content and texture (Table 1), which in turn are governed by the degree of separation and rate of cooling of the magma. The most common crystalline rocks are granites and metamorphics, which form most of the interiors of the continents. The texture of the granitic rocks is controlled by the rate of cooling, with the slowest rates providing the largest crystals and the more rapid rates resulting in smaller grains. Most igneous rocks are aggregates of two or more minerals. Granite, for example, is composed almost entirely of quartz and feldspar with a minor constituent of other minerals. Granitic rocks formed deep within the crust, where crystal growth was controlled by the cooling rate of the magma and the available space.

Large crystals probably formed late in the crystallization of a magma body such as a batholith, which provides new additions to the crust. The

TABLE 1 COMMON IGNEOUS ROCKS

	Felsic	Intermediate	Mafic	Ultramafic
Intrusive	Granite	Diorite	Gabbro	Peridotite
Extrusive	Rhyolite	Andesite	Basalt	None
Mineral composition	Quartz	Hornblend	Calcium feldspar	Olivine
	Potassium feldspar	Sodium feldspar	Pyroxenes	Pyroxines
		Calcium feldspar		
Minor mineral constituents	Sodium feldspar	Biotite	Olivine	Calcium feldspar
	Muscovite	Pyroxenes	Hornblend	
	Biotite			
	Hornblend			

best-known of these include the Sierra Nevada batholith (Fig. 14), the California batholith, and the Andean batholith. Large crystals also form in the presence of volatiles such as water and carbon dioxide, permitting them to grow in a smaller volume. As a magma body slowly cools, possibly over a period of a million years or more, the crystals grow directly out of the fluid melt or out of the volatile magmatic fluids that invaded the surrounding rocks.

Igneous and sedimentary rocks subjected to the intense temperatures and pressures of the Earth's interior, by heat generated near magma bodies, by shear pressures from earth movements, or by strong chemical reactions that do not result in melting produce metamorphic rocks. Metamorphism causes dramatic changes in texture, mineral composition, or both. It produces new textures by recrystallization, which causes minerals to grow into larger crystals. New minerals are also created by recombining chemical elements to form new associations. Water and gasses from nearby magma bodies also aid in the chemical changes that take place in rocks by conveying chemical elements from one place to another.

Heat is the primary agent for recrystallization, and often deep burial is required to generate the temperatures and pressures required for extensive metamorphism. Varying degrees of metamorphism are also achieved at shallower depths in geologically active areas with higher thermal gradients, where the temperature increases with depth much faster than normal. During metamorphism, rocks behave plastically and can bend or stretch due to the high temperatures and pressures from the overlying strata. Consequently, metamorphic rocks make up the largest constituent of the Earth's crust.

Figure 14 *Sheet joints formed in granitic rocks of the Sierra Nevada batholith, California.*

(Photo by N. K. Huber, courtesy USGS)

CONTINENTAL CRUST

The Earth has the thinnest crust of all the terrestrial planets (Fig. 15); even the moon's crust is thicker. The crust represents less than 1 percent of the Earth's radius and 0.3 percent of its mass. Including continental margins and small shallow regions in the ocean, continental crust covers about 45 percent of the Earth's surface (Tables 2 & 3), with an average thickness of about 28 miles and an average height of 2.8 miles above the seafloor. The crust is built like a layer cake with sedimentary rocks on top, granitic and metamorphic rocks in the middle, and basaltic rocks on the bottom. This gives the crust a structure similar to a jelly sandwich with a pliable middle layer sandwiched between a hard upper crust and a rigid lithosphere.

The Earth did not develop a permanent crust until around 4 billion years ago after a period of massive meteorite bombardment between 4.2 and 3.8 billion years ago. Consequently, the oldest rocks on the Earth's surface are not nearly as old as the planet itself. During the first half billion years or so, the Earth was in a fiery turmoil, and any rocks solidifying during this period soon remelted. The meteorite bombardment also melted the crust by impact friction. Thousands of 50-mile-wide meteorites pounded the Earth and converted 30 to 50 percent of the crust into impact basins, which later filled with water when the first rains came.

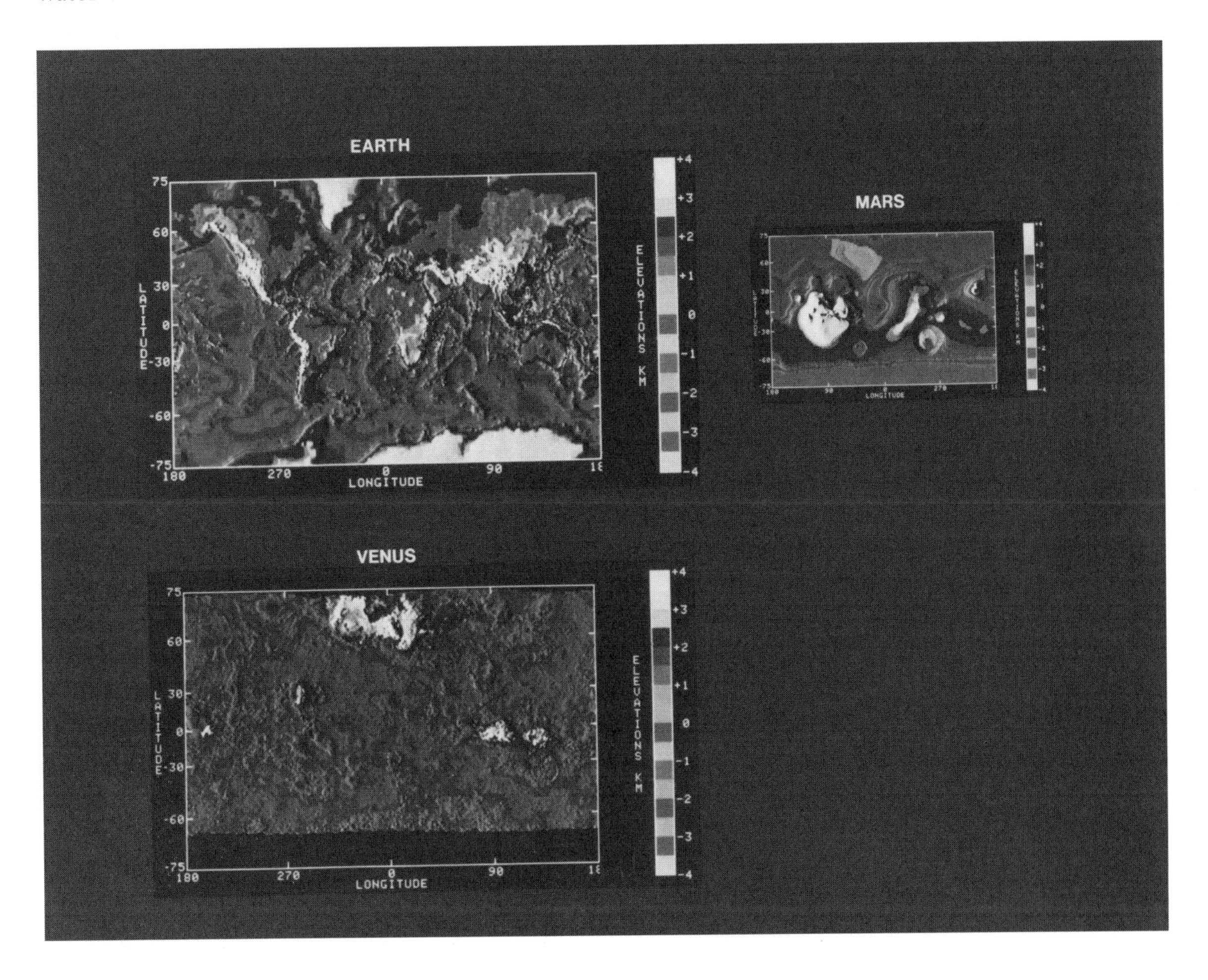

Figure 15 *Comparison of topographies of Earth, Mars, and Venus.*

(Photo courtesy NASA)

TABLE 2 CLASSIFICATION OF THE EARTH'S CRUST

Environment	Crust Type	Tectonic Character	Thickness (Miles)	Geologic Features
Continental crust overlying stable mantle	Shield	Very stable	22	Little or no sediment, exposed Precambrian rocks
	Midcontinent	Stable	24	
	Basin & Range	Very unstable	20	Recent normal faulting, volcanism, and intrusion; high mean elevation
Continental crust overlying unstable mantle	Alpine	Very unstable	34	Rapid recent uplift, relatively recent intrusion; high mean elevation
	Island arc	Very unstable	20	High volcanism, intense folding and faulting
Oceanic crust overlying stable mantle	Ocean basin	Very stable	7	Very thin sediments overlying basalts, no thick Paleozoic sediments
Oceanic crust overlying unstable mantle	Ocean ridge	Unstable	6	Active basaltic volcanism, little or no sediment

Around remote lakes and tundra of northwest Canada is metamorphic granite called Acasta Gneiss dating about 4.2 billion years old, making it the oldest terrestrial crust on Earth. It lies within a large protocontinent known as the Slave craton, suggesting that small tectonic plates crashed into each other as early as 4 billion years ago. Since the rocks are composed of granite, they indicate that the Earth was well under way forming continental crust by this time. Only about 5 to 8 percent of the present continental crust was in existence between 4 and 3 billion years ago. However, by 2.7 billion years ago, most of the present terrestrial crust had formed, indicating highly rapid continental growth.

Rocks found in southwest Greenland dating around 3.8 billion years old are composed of metamorphosed sediments that were originally laid down in a marine environment, signifying the Earth had a significant ocean by this time. Older rocks have been found in Antarctica and Africa, but for the most part few rocks date beyond 3.7 billion years old. This suggests that no major continents were in existence by this time, but only thin slices of crust that wandered across the face of the planet driven by rapid convective motions in the mantle.

As convection currents began to slow due to the loss of internal heat, lighter rock materials migrated toward the surface to form a basaltic scum. In essence, the crust is composed of waste products of the mantle. The reworking of this primitive crust as it was subducted into the mantle and remelted formed the first granitic rocks. The bulk of the crust is composed of oxygen, silica, and aluminum, which form the granitic and metamorphic rocks in the cores of the continents.

Sediments thrust deep into the mantle were subjected to intense internal heat. The rocks either changed crystal structure or melted entirely and became magma. The buoyant magma rose to the surface in blobs called diapirs from Greek meaning "to pierce." If the magma broke through the surface, it produced volcanic eruptions. Otherwise it remained buried in the crust, forming large granitic bodies called plutons. The original continents were free-roaming slivers of crust constantly colliding with each other. As the Earth continued to cool, their erratic wanderings slowed and they began to stick together, forming over a dozen protocontinents.

Eventually, all the protocontinents combined into larger continents. Most of the continental crust was created when lithospheric plates collided. The colli-

TABLE 3 COMPOSITION OF THE EARTH'S CRUST

Crust Type	Shell	Average Thickness (in Miles)	Percent Composition of Oxides						
			Silica	Alum	Iron	Magn	Calc	Sodi	Potas
Continental	Sedimentary	2.1	50	13	6	3	12	2	2
	Granitic	12.5	64	15	5	2	4	3	3
	Basaltic	12.5	58	16	8	4	6	3	3
Total		27.1							
Subcontinental	Sedimentary	1.8							
	Granitic	5.6				Same as above			
	Basaltic	7.3							
Total		14.7							
Oceanic	Sedimentary	0.3	41	11	6	3	17	1	2
	Volcanic sedimentary	0.7	46	14	7	5	14	2	1
	Basaltic	3.5	50	17	8	7	12	3	<1
Total		4.5							
Average		15.4	52	14	7	4	11	2	

sions also greatly deformed the crust over a broad area. Worldwide continental-forming events took place roughly 2.9 to 2.6 billion years ago, 1.9 to 1.7 billion years ago, 1.1 to 0.9 billion years ago, and during the past 600 million years.

OCEANIC CRUST

Oceanic crust is much thinner than continental crust, and in most places it is only 3 to 5 miles thick. The oceanic crust is remarkable for its consistent thickness and temperature, averaging about 4 miles thick and varying no more than 20 degrees Celsius over most of the globe. By comparison, the continental crust averages 25 to 30 miles in thickness, and at mountain ranges, it reaches depths of 45 miles. Continental crust has an average density of 2.7 times the density of water, whereas oceanic crust is 3.0. Because the mantle's density is 3.4, the continental and oceanic crust remain afloat above the mantle.

Like icebergs, most of the crust resides beneath the surface. The long-lived continental roots that underlie mountain ranges can extend downward as much as 250 miles into the upper mantle. Continental crust is also 20 times older than oceanic crust, which is no older than 170 million years because oceanic crust is recycled back into the mantle. Most of the seafloor has since disappeared into the Earth's interior to provide the raw materials for the continued growth of the continents.

Oceanic crust does not form as a single homogeneous mass, but comprises long narrow ribbons laid side by side with fracture zones between. The oceanic crust is like a layer cake with three distinct strata. It has an upper layer of pillow basalts, formed when lava extruded undersea at great depths; a middle layer of a sheeted-dike complex, consisting of a tangled mass of feeders that brought magma to the surface; and a lower layer of gabbros, which are coarse-grained rocks that crystallized slowly under high pressure in a deep magma chamber. Gabbros with higher amounts of silica solidify out of the basaltic melt and accumulate in the lower layer of the oceanic crust.

New oceanic crust forms at spreading ridges, where basalt oozes out of the mantle through rifts on the ocean floor, generating about 5 cubic miles of new oceanic crust every year. The mantle below the spreading centers, where new oceanic crust is created, consists mostly of peridotite, a strong, dense rock composed of iron and magnesium silicates. As the peridotite melts on its 30- to 40-mile journey to the base of the crust, part of it becomes highly fluid basalt, the most common magma erupted on the Earth's surface.

Most of this volcanism occurs on the ocean floor at spreading centers (Fig. 16), where the oceanic crust is being pulled apart. Some molten magma erupts as lava on the surface of the ridge through a system of vertical passages. Once at the surface, the liquid rock flows down the ridge and hardens into

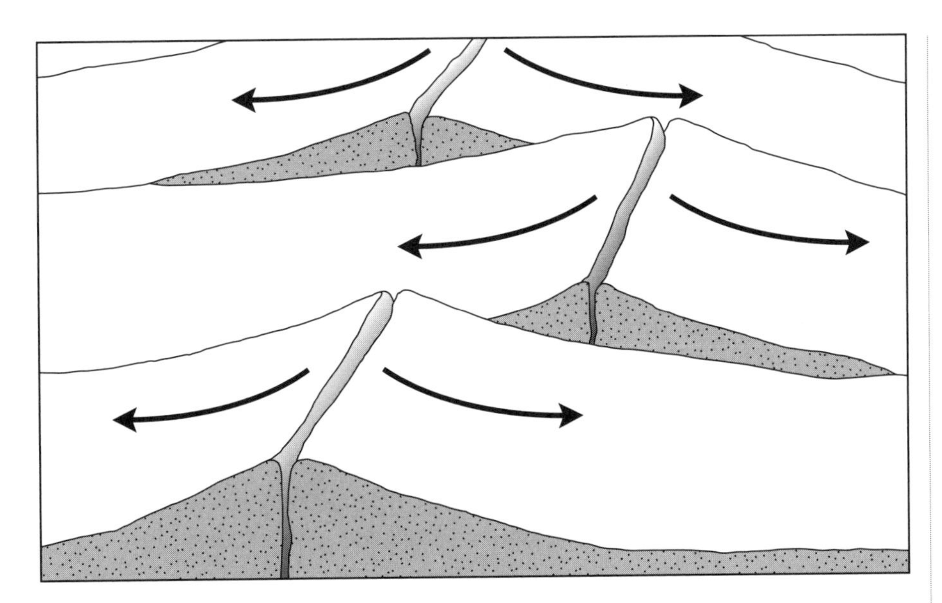

Figure 16 *Spreading centers spread segments of the seafloor apart.*

sheets or rounded forms of pillow lavas, depending on the rate of extrusion and the slope of the ridge. Magma rising from the upper mantle extrudes onto the ocean floor and bonds to the edges of separating plates.

Much of the magma solidifies within the conduits above the magma chamber, forming massive vertical sheets called dikes that resemble a deck of cards standing on end. Individual dikes measure about 10 feet thick, about 1 mile wide, and extend horizontally about 3 miles. Periodically, lava overflows onto the ocean floor in gigantic eruptions, providing several square miles of new oceanic crust yearly. As the oceanic crust cools and hardens, it contracts, forming fractures through which water circulates.

Oceanic crust begins relatively thin and eventually thickens by the underplating of new lithosphere from the upper mantle and the accumulation of overlying sediment layers. As an oceanic plate ages, it reaches a thickness of about 60 miles after about 60 million years. By the time the oceanic crust spreads out as wide as the Atlantic Ocean, the portion near continental margins where the ocean is the deepest is more than 50 miles thick. Eventually, the oceanic crust becomes too thick and heavy to remain on the surface and subducts into the mantle, where it melts to provide material for new crust.

When a continental plate collides with an oceanic plate, the latter bends downward and subducts into the mantle. The oceanic plate remelts in the Earth's interior, acquires new minerals from the mantle, and reemerges at volcanic spreading centers, most of which are in the ocean. This rejuvenates the ocean crust while spreading apart the continents, which ride on thick slabs of rock.

The Earth's outer shell is fashioned out of eight major and about a half dozen minor tectonic plates. The plates are composed of the upper brittle crust and the upper brittle mantle together called the lithosphere, which averages about 60 miles thick. The lithosphere consists of the rigid outer layer of the mantle and the overlying continental or oceanic crust and rides on the semimolten layer of the upper mantle or asthenosphere. This structure is important for the operation of plate tectonics, without which the crust would become jumbled up slabs of rock like Arctic pack ice.

The lithospheric plates act as rafts riding on a sea of molten rock and carry the crust around the surface of the globe. The plates diverge at spreading ridges and converge at subduction zones, which are depicted on the ocean floor as deep-sea trenches, where the plates dive into the mantle and remelt. The plates and oceanic crust are continuously recycled through the mantle, but the continental crust due to its greater buoyancy remains mostly on the surface.

After learning about the outer layer of the Earth's rocks and their interactions with each other, the next chapter will examine the erosional and depositional processes occurring on the continents and in the oceans.

2

EROSION AND SEDIMENTATION

THE BUILDING OF LANDFORMS

This chapter discusses the weathering, erosional, and sedimentary processes that create landforms. Erosion is a natural geologic process that cuts down tall mountains and carves deep canyons and has been doing so since the very beginning of time. No matter how pervasive is the formation of mountain ranges by the forces of uplift, they eventually lose the battle with erosion and are worn to the level of the prevailing plain. Erosion also gouges deep ravines in the hardest rock and has obliterated most geologic structures, including structures build by ancient civilizations.

Most of the Earth's surface is covered by a thin veneer of sediment, and sedimentary rocks are encountered more frequently than any other type. They not only present impressive scenery from ragged mountains to jagged canyons, but also much of the wealth of the world, including valuable ores and petroleum. The sedimentary environment provides the conditions necessary for the formation of fossils, which give important clues about the history of the Earth. The constant shifting of sediments on land and the accumulation of deposits in the ocean assures that the face of the Earth will continue to change with time.

EFFECTS OF EROSION

Erosion rates vary depending on precipitation, the topography of the land, the type of rock and soil material, and the amount of vegetative cover. Yearly, the Mississippi River dumps more than 250 million tons of sediment into the Gulf of Mexico, widening the Mississippi Delta and slowly building up Louisiana and nearby coastal regions. The Gulf Coast states from East Texas to the Florida Panhandle were built by sediments eroded from the interior of the continent and hauled in by the Mississippi and other rivers. The Imperial Valley of southern California owes its rich soil to the Colorado River, which carved out the mile-deep Grand Canyon and deposited its sediments 3 miles thick in the region.

The process of erosion is delicately balanced by crustal buoyancy, which keeps the continents afloat. Therefore, erosion can only shave off the top portion of the continental crust before its mean height falls below sea level, at which point erosion ceases and sedimentation commences. In the past, erosion rates were probably higher than today and the relief of the land was not nearly as great. Eons of mountain building and erosion were required to create the present landscape of tall mountains and deep canyons.

The rise of active mountain chains such as the Himalayas (Fig. 17) is matched by erosion so that their net growth is practically zero. Water and wind gradually erase most signs of once splendid mountain ranges. The cores of the world's mountains contain ancient rocks that were once buried deep in the bowels of the Earth and are now thrust high above. Huge blocks of granite that formed the interiors of the continents were pushed up by tectonic forces operating deep within the Earth and exposed by erosion.

The shaping of mountains depends as much on the destructive forces of erosion as on the constructive power of plate tectonics. The interactions between tectonic, climatic, and erosional processes exert strong control over the shape and height of mountains as well as time needed to build or destroy them. Erosion might actually be the most powerful agent of mountain building by removing mass restored by isostasy, which lifts the entire mountain range to replace the missing mass. If the rate of erosion matches the rate of uplift, the size and shape of mountains can remain stable for upward of millions of years. As the mountains age, however, the crust supporting them thins out, and erosion takes over to bring down even the most imposing ranges.

Soil erosion causes the most widespread degradation of the land surface. Falling rain erodes surface material by impact and runoff. The impact of raindrops striking the ground with a high velocity loosens material and splashes it up into the air. On hillsides, some of this material falls back at a point lower down the slope. About 90 percent of the energy is dissipated by the splash

Figure 17 *View of the Himalaya Mountains of India and China from the space shuttle.*

(Photo courtesy NASA)

impact. Most of the impact splashes are to a height of up to 1 foot, and the lateral splash movement is about four times the height.

Impact erosion is most effective in regions with little or no vegetative cover and subject to sudden downpours, such as desert areas. Splash erosion accounts for the puzzling removal of soil from hilltops where little runoff occurs. It also can ruin soil by splashing up the light clay particles, which are carried away by runoff, leaving behind infertile silt and sand. The degree of erosion depends on the slope steepness and the type and amount of vegetative cover. Rainwater not infiltrating into the ground runs down the hillside and erodes the soil, cutting deep gullies into the terrain (Fig. 18).

Figure 18 *Severe soil erosion on farmland near Viola, Idaho, showing gully and rills.*

(Photo by Carrol Tyler, courtesy USDA Soil Conservation Service)

DRAINAGE PATTERNS

About 3.5 million miles of rivers and streams cross the United States, and many parts of the nation are prone to flooding. Flash floods are the most intense form of flooding. They are local floods of great volume and short duration and are generally caused by torrential rains or cloudbursts associated with severe thunderstorms on a relatively small drainage area. Flash floods also result from a dam break or from a sudden breakup of an ice jam, causing the release of a large volume of flow in a short time.

A special type of flash flood occurred during the 1980 eruption of Mount St. Helens, which created major mudflows and flooding (Fig. 19) from melted glaciers and snow on the volcano's flanks. Flood-hazard zones extend considerable distances down some valleys. Flood zones of the volcanoes in the western Cascade Range, for example, can reach as far as the Pacific Ocean.

Flash floods can take place in almost any part of the country. Violent thunderstorms can produce flash floods on widely dispersed streams, resulting in high flood waves. The discharges quickly reach a maximum and diminish almost as rapidly. Floodwaters frequently contain large quantities of sediment and debris collected as they sweep clean the stream channel. Flash floods are especially common in the mountainous areas and desert regions of the American West. They are particularly dangerous in areas where the terrain is steep, surface runoff rates are high, streams flow in narrow canyons, and severe thunderstorms are commonplace.

Riverine floods are produced by precipitation over large areas, by the melting of the winter's accumulation of snow, or both. They differ from flash floods in both extent and duration and take place in river systems whose tributaries drain large geographic areas and encompass many independent river basins. Floods on large river systems might last from a few hours to many days. The flooding is influenced primarily by variations in the intensity and the amount and distribution of precipitation. Other factors directly affecting flood runoff include the condition of the ground, the amount of soil moisture, and the vegetative cover.

River channel storage, changing channel capacity, and timing of flood waves, which depend on the river size, control the movement of floodwaters. As the flood moves down the river system, temporary storage in the channel reduces the flood peak. As tributaries enter the main stream, the river increas-

Figure 19 *The May 18, 1980, eruption of Mount St. Helens caused extensive flooding and sedimentation along the Cowlitz River, Cowlitz County, Washington.* (Photo courtesy USGS)

Figure 20 *The northern foothills and Arctic coastal plain province east of the Kukowruk River, Utukok-Corwin region, Northern Alaska.*

(Photo by R. M. Chapman, courtesy USGS)

es in size farther downstream. Since tributaries are not the same size or spaced uniformly, their flood peaks reach the main stream at different times, thereby smoothing out the peaks as the flood wave moves downstream.

Most water precipitating on the continents is lost through floods or is held in lakes, swamps, and soil. About a third is base flow, which is the stable runoff of rivers and streams. Another third is subsurface flow, which discharges mostly through evaporation, and only about 1 percent reaches the ocean.

Groundwater travels very slowly and can make continental-scale journeys lasting up to millions of years.

A drainage basin is the entire area from which a stream and its tributaries receive their water. For example, the Mississippi River and its tributaries drain a tremendous section of the central United States, reaching from the Rockies to the Appalachians. Moreover, each of its tributaries has its own drainage area, which forms a part of the larger basin. Individual streams and their valleys are joined into networks, which display various types of drainage patterns, depending on the terrain (Fig. 20).

Drainage patterns might be dendrite, resembling the branches of a tree, if the terrain is of uniform composition, or trellis, a rectangular pattern due to differences in the bedrock's resistance to erosion. Rectangular drainage patterns also occur if the bedrock is crisscrossed by fractures, which form zones of weakness that are particularly susceptible to erosion. Streams that radiate outward in all directions from a topographic high, such as a volcano or dome, form radial stream patterns.

Stream drainage patterns are influenced by topographic relief and rock type. They provide important clues about the type of geologic structure in an area. In addition, the color and texture of the structure carry information about the rock formations that it comprises. Surface expressions such as anticlines, synclines, folds, and domes (Fig. 21) bear clues about the subsurface structure. Various types of drainage patterns infer variations in the surface

Figure 21 *A dome of redbeds at Bellvue, Colorado.*

(Photo by W. T. Lee, courtesy USGS)

lithology. The drainage pattern density is another good indicator of the lithology. Variations in the drainage density are also associated with changes in the coarseness of the alluvium.

In areas of exposed bedrock, these patterns depend on the lithologic character of the underlying rocks, the attitude of these rock bodies, and the arrangement and spacing of planes of weakness encountered by runoff. Any abrupt changes in the drainage patterns are particularly important because they signify the boundary between two rock types, which might be good locations to explore for minerals.

EROSIONAL FEATURES

The most impressive geologic features the planet has to offer were carved out of the crust by erosional processes. Ancient geologic structures have long been erased by active erosional agents. Nature's hydrologic cycle, involving the flow of water from the ocean, across the land, and back to the sea, provides the most effective forces of erosion. Massive glaciers carved out some of the most monumental landforms, while outwash streams from glacial meltwater further eroded the landscape. Areas lacking rainfall or snowfall such as deserts and tundra retain much of their geologic structures simply because they experience little erosional activity.

Unconformities are common and seen most everywhere. They are among the most remarkable of geologic features and reflect processes that represent the entire spectrum of geologic events. Unconformities contain pockets of time or gaps in the geologic record where no rocks exist. Often, the older underlying rocks are folded or tilted and overlain by younger flat-lying sediments, producing bewildering patterns of discordant rocks and angular contacts.

An unconformity is created when layers of sediments build up on an area previously eroded by water and wind. The presence of an unconformity indicates a land surface reaching above sea level. The oldest evidence of dry land is a 3.5-billion-year-old unconformity in northwestern Australia in the geologic province known as the Pilbara craton. The Australian rocks appear to contain paleosol, or fossilized soil, suggesting active weathering agents on the Earth's surface at this time. Surprisingly, ancient fossil cells, the oldest solid evidence of life, were found in the very rock formation that lies directly above the unconformity, known as the Warrawoona Group.

Massive sandstone cliffs found in the western United States (Fig. 22) have been slowly eroding for millions of years. Perhaps nowhere else on Earth is this process more illuminating than in the Grand Canyon of northern Arizona, which contains an imposing unconformity on the canyon walls. The Grand Canyon was carved out by the roaring Colorado River, now merely a trickle of

Figure 22 Cretaceous Mancos shale and Mesaverde formations, Mesa Verde National Park, Colorado.

(Photo by L. C. Huff, courtesy USGS)

its former self. Farther north is Bryce Canyon National Park in southern Utah, where fantastic pillars were carved out of the colorful Wasatch Formation. Similarly colored sediments are responsible for the Painted Desert of Arizona and the Badlands of South Dakota (Fig. 23), where short, steep slopes were eroded by a myriad of small streams, forming a unique drainage network.

Erosional processes taken to extreme have created Monument Valley on the border between Utah and Arizona. Isolated or groups of monuments rise 1,000 feet or more off the desert floor. A resistant cap rock preserved the sediments below, while the rest of the landscape eroded away. A similar situation only on a broader scale exists in flattop mountains, where a remnant of the original peneplain, literally meaning almost a plain, is protected by a more resistant layer of sandstone. Many mesas, including the Grand Mesa in western Colorado, the largest in the world, owe their existence to an upper layer of resistant basalt.

Dikes are formed by tabular magma bodies occupying a crack or a fissure in the crust. They are usually harder than the surrounding material and produce long ridges when exposed by erosion. One of the best examples of this feature is Shiprock, New Mexico, where large dikes radiate outward from a 1,400-foot-tall volcanic neck, left standing when the overlying sediments eroded. Another good example of this type of structure is Devils Tower in Wyoming, composed of solidified magma that filled a volcanic pipe. Erosion has left the more resistant rock standing high above surrounding terrain.

Volcanic calderas are huge gaping pits in the crust formed by the collapse of massive volcanoes. Calderas are widened even further by the process of erosion to the point where they are sometimes difficult to observe on the ground. However, many such calderas are easily spotted from aircraft or satel-

Figure 23 *Rugged outcrops of the Wasatch Formation, Badlands National Park, South Dakota.*

(Photo courtesy National Park Service)

lites. This is also true of large meteorite craters, which might be so severely eroded the only way they can be detected is from above.

SEDIMENTARY PROCESSES

Sediments are derived from the weathering of the crust, and most sedimentary processes take place very slowly on the ocean floor. Marine sediments consist of material washed off the continents. Therefore, most sedimentary rocks form along continental margins and in the basins of inland seas. One such sea invaded the interior of North America during the Jurassic and Cretaceous periods (Fig. 24). Areas with high sedimentation rates form deposits thousands of feet thick. Where these deposits are exposed on the surface, individual sedimentary beds can be traced hundreds of miles.

The formation of sedimentary rock begins with erosion. The continents are the primary sites of erosion, whereas the oceans are the principal sites of sedimentation. Rocks weather by several processes, including the action of rain, wind, and ice. Loose sediment grains are carried downstream to the ocean. Rivers such as the Amazon and the Mississippi transport enormous quantities of sediment derived from the interiors of their respective continents. The towering landform created by the collision of India and Asia is the greatest single source of sediment today. The sediment is hauled out by major rivers draining the region and emptied into the Bay of Bengal. This accounts for 40 percent of the total amount of sediment discharged into the ocean by all the rivers in the world.

Each year, an estimated 25 billion tons of sediment are carried by stream runoff into the ocean, where it settles onto the continental shelf. The continental shelf extends to 100 miles or more and reaches a depth of roughly 600 feet. In most places, the continental shelf is nearly flat with an average slope of only about 10 feet per mile, comparable to the slopes of many coastal regions. Beyond the continental shelf lies the continental slope, which extends to an

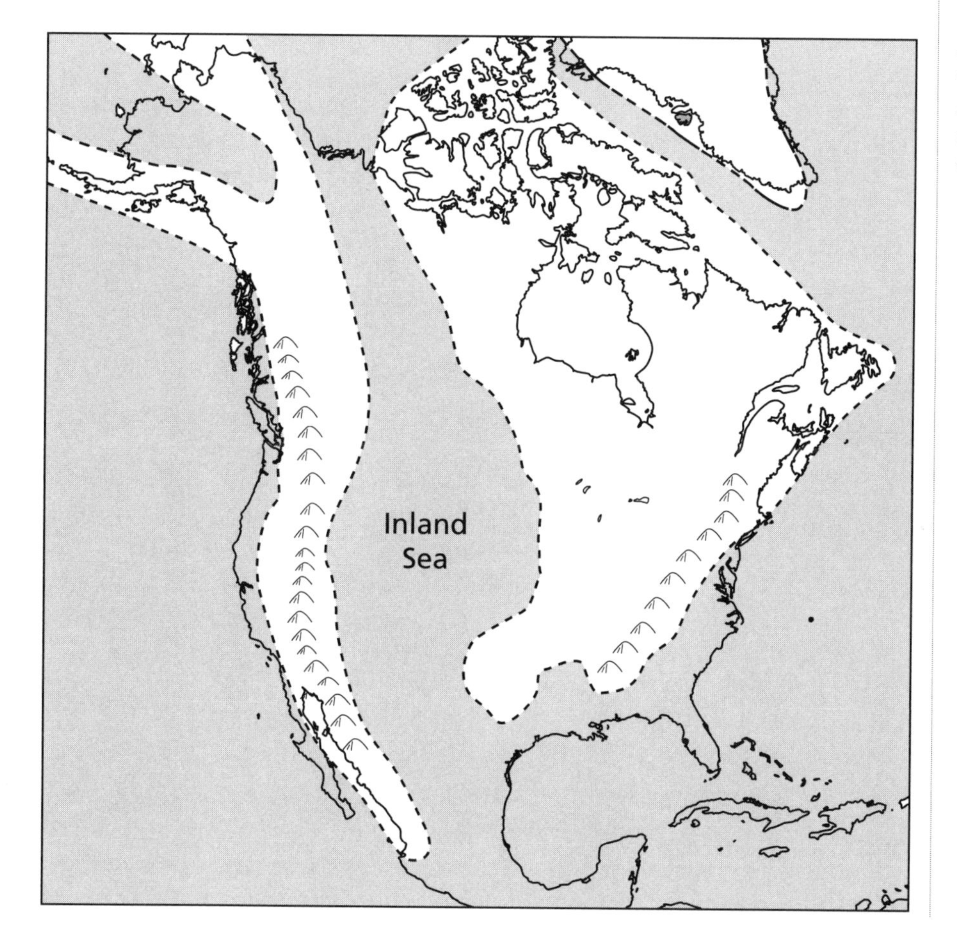

Figure 24 *The Cretaceous interior sea of North America where thick deposits of sediments were laid down.*

average depth of more than 2 miles. It has a steep angle of several degrees, comparable to the slopes of many mountain ranges. Sediments reaching the edge of the continental shelf slide down the continental slope under the influence of gravity. Often, huge masses of sediment cascade down the continental slope by gravity slides, sometimes gouging out steep submarine canyons.

Most minerals in sedimentary rocks precipitated directly from seawater. When land is eroded, some 3 billion tons of rock are dissolved in water and carried by streams to the sea annually. This is sufficient to lower the entire land surface of the Earth by as much as an inch in only 2,000 years. It is also one reason the ocean is so salty. Besides ordinary table salt, seawater contains large amounts of calcium carbonate, calcium sulfate, and silica, which precipitate from seawater by chemical and biologic processes.

In dry regions where dust storms are common, loose sediment is carried by the wind. Windblown sediments landing in the ocean slowly build deposits of abyssal red clay, whose color signifies its terrestrial origin. However, most windblown sediments remain on the continents. They often accumulate into thick deposits of loess, which is distinguishable from other sediments by its thin, uniform bedding. Most loess deposits in the central portion of the United States were laid down during the Pleistocene ice ages. During glaciation, regions not covered by ice dried out, causing widespread desertification.

Fluvial deposits are riverborne sediments that remain on the continents after erosion. When a river clogs with sediments and fills its channel, it spills over onto the adjacent floodplain and carves out a new river course. This forces the river to meander along, forming thick sediment deposits in broad floodplains that can fill an entire valley. Fluvial deposits are recognized on outcrops by their coarse sediment grains and cross-bedding features (Fig. 25), created when the stream meandered back and forth over old river channels.

Floodwaters rapidly flowing out of dry mountain regions carry a heavy sediment load, sometimes including blocks the size of automobiles. When the stream reaches the desert, the water rapidly percolates into the desert floor, causing the flood to halt suddenly. After erosion has peeled away the outer layers of mud, huge monoliths are left standing in the middle of nowhere as a testament to the immense power of water in motion.

SEDIMENTARY ROCKS

Sedimentary rocks are classified into clastic, carbonate, and evaporate deposits. Clastic sedimentary rocks are composed mainly of fragments broken loose from a parent material, deposited by mechanical transport, and cemented into hard rock. The sediments are usually derived from the weathering or decomposition of igneous, metamorphic, and other sedimentary rocks. The rocks

Figure 25 *Cross-bedded Navajo sandstone, Johnson Canyon, Kane County, Utah.*

(Photo by W. H. Jackson, courtesy USGS)

were weathered, or broken down, into sediment grains by the action of water, wind, cycles of heating and freezing, and plant and animal activity. Weathering breaks rocks apart or causes their outer layers to peel or spall off in a process known as exfoliation.

The products of weathering include a wide range of materials from very fine-grained sediments to large boulders. Erosion, either by rain, wind, or glacial ice, eventually brings the sediments to streams and rivers, which in turn empty into the ocean. Angular sediment grains indicate a short time spent in transit, whereas rounded sediment grains indicate severe abrasion from long-distance travel or from reworking by fast-flowing streams or pounding ocean waves.

Solid rock exposed on the surface is chemically broken down into clays and carbonates and mechanically broken down into silts, sands, and gravels. Streams, heavily laden with sediments, overflow their beds, forcing them to take several detours as they meander toward the sea. When the streams reach the ocean, their velocity falls off sharply, and their sediment load drops out of suspension. Meanwhile, chemical solutions carried by the rivers are thoroughly mixed with seawater by ocean waves and currents.

When riverborne sediments reach the ocean, they settle out of suspension according to grain size. The course-grained sediments settle out near the turbulent shore, and the fine-grained sediments settle out in calmer waters farther out to sea. As the shoreline advances toward the sea from the buildup of

Figure 26 *Interbedded mudstone and sandstone formation, Capitol Reef National Monument, Utah.*

(Photo by J. R. Stacy, courtesy USGS)

coastal sediments or by a falling sea level, finer sediments are progressively covered over by courser ones. As the shoreline recedes inland from the lowering of the land surface or by a rising sea level, courser sediments are covered over by progressively finer ones. This produces a recurring sequence of sandstones, siltstones, and shales (Fig. 26). Gravels are rare in the ocean and are mainly transported from the coast to the deep abyssal plains by submarine slides called turbidity currents.

As the weight of the overlying sedimentary layers presses downward on the lower strata, the sediments are lithified into solid rock, providing a geologic column of alternating beds of limestone, shales, siltstones, and sandstones. If these rocks are subjected to the heat and pressure of the Earth's interior, they are metamorphosed consecutively into marble, slate, quartzite, and schist.

Shales and mudstones are the most abundant sedimentary rocks because they are the main weathering products of feldspars, the most abundant minerals.

Furthermore, all rocks are eventually ground down to clay-size particles by abrasion. Because clay particles are so small and sink so slowly, they normally settle out in calm, deep waters far from shore. Compaction by the weight of the overlying sediments squeezes out water between sediment grains, lithifying the clay into mudstone if massive or into shale if fissile, or thinly bedded.

Clastic sedimentary rocks are classified according to grain size. Gravel-size sediments are called conglomerates if rounded and breccias if angular. Conglomerates are composed of abundant quartz and chalcedony such as flint or chert. Breccias are relatively rare and are indicative of terrestrial mudflows or submarine slides. Debris flows piled up on the continental slope can produce a coarse carbonate rubble known as brecciola. Volcanic breccias, also known as agglomerates, are consolidated pyroclastic fragments. Moraines and tillites are glacial deposits composed of boulders and gravel-size sediments.

Sandstones are composed mostly of quartz grains roughly the size of beach sands. If sandstone contains abundant feldspar, it is called an arkose. Graywacke, sometimes called dirty sandstone, is a dark, coarse-grained sandstone with a clay matrix and is believed to be deposited by submarine turbidity currents. Siltstones are composed of fine quartz grains just visible to the naked eye. Shales and mudstones are composed of the finest sedimentary particles, whose grains are invisible to the unaided eye.

Clastic sediments are lithified into solid rock mainly by compaction for fine-grained sediments and cementation for coarse-grained sediments. With increasing weight from the overlying sedimentary layers, individual grains are compacted and solidified. Minerals such as calcium carbonate, silica, or iron oxide are deposited between coarse sediment grains and are used as cementing agents.

Coal beds (Fig. 27) are considered sedimentary rock, even though they are not derived from clastic sediments or chemical precipitation. Coal originated from compacted plant material that once grew in lush swamps. Often between easily separated layers of coal or associated fine-grained sedimentary beds are carbonized remains of ancient plant stems and leaves. Black or carbonized shales also originated in the ancient coal swamps, and traces of plant life can be found between easily split shale layers.

Non-clastic, or precipitate, rocks are formed by biologic or chemical precipitation of minerals dissolved in water. The term *precipitate* is actually a historical misnomer carried over from the days when these rocks were thought to form similar to the precipitation of ice crystals. Rainwater contains a small amount of carbonic acid from the chemical reaction of atmospheric water vapor and carbon dioxide. The carbonic acid dissolves calcium and silica minerals in surface rocks to form bicarbonates and silicates. These are transported to the ocean by rivers and are thoroughly mixed with seawater by waves and currents.

The bicarbonates precipitate out of seawater by direct chemical processes or by biologic activity, the most common means. Organisms use

Figure 27 *A thick coal bed, Sunnyside mine, Carbon County, Utah.*

(Photo by D. J. Fisher, courtesy USGS)

calcium bicarbonate to build supporting structures such as shells composed of calcium carbonate. When the organisms die, their shells fall to the bottom of the ocean, where thick deposits of calcium carbonate slowly build up to form limestone.

Limestone is the most common precipitate rock and is mostly produced by biologic activity. This is evidenced by abundant fossilized marine life in limestone beds. Some limestones are chemically precipitated directly from seawater, and a minor amount is precipitated in evaporate deposits. Dolomite resembles limestone, but is created when calcium in limestone is replaced by magnesium. It is more resistant to erosion from acid rain, which explains why the Dolomite Alps of Europe remain among the most impressive mountain ranges in the world. On the other extreme, chalk is a soft, porous calcium carbonate rock, not to be confused with the chalk used on classroom blackboards, which is instead composed of calcium sulfate.

Most limestones originated in the ocean, and some thin limestone beds were deposited in lakes and swamps. Limestones constitute approximately 10 percent of all exposed sedimentary rocks. Many limestones form massive formations (Fig. 28). They are recognized by their typically light gray or light brown color. Limestones are among the best suited rocks for fossil preservation because of the nature of their sedimentation, often involving shells and skeletons of dead marine life buried and fused into solid rock.

Whole or partial fossils comprise many limestones, depending on whether they were deposited in quiet or agitated waters. Tiny, spherical grains called oolites are characteristic of agitated water, whereas lithified layers of limey mud called micrite are characteristic of calm waters. In quiet waters, undisturbed by waves and currents, whole organisms with hard body parts are buried in the calcium carbonate sediments and are later lithified into limestone.

Carbonate sediments were deposited in shallow waters, probably less than 50 feet deep, and mainly in intertidal zones, where marine organisms were plentiful. Coral reefs, formed in shallow water where sunlight can easily penetrate for photosynthesis, contain abundant organic remains. Many ancient reefs are composed largely of carbonate mud that contains larger skeletal remains.

Most carbonate rocks began as sandy or muddy calcium carbonate material. The sand-size particles are composed of fragmented skeletal remains of invertebrates and shells of calcareous algae that rain down from above. The skeletal remains might have been broken up by mechanical means, such as the pounding of the surf, or by the activity of living organisms. Further breakdown into dust-size particles produces a carbonate mud. It is the most common constituent of carbonate rocks and forms a matrix known as micrite. Under certain conditions, the carbonate mud dissolves in seawater and is redeposited elsewhere on the ocean floor, forming a calcite ooze that later lithifies into limestone.

As calcareous sediments accumulate into thick deposits on the ocean floor, deep burial of the lower strata produces high pressures, which lithifies the beds into carbonate rock, consisting mostly of limestone or dolomite. If

Figure 28 *The eastern Sawtooth Range, Lewis and Clark County, Montana.*

(Photo by M. R. Mudge, courtesy USGS)

fine-grained calcareous sediments are not strongly lithified, they form deposits of soft, porous chalk. Limestones typically develop a secondary crystalline texture, resulting from the growth of calcite crystals by solution and recrystallization following the formation of the original rock.

Some carbonate rocks were deposited in deep seas. The maximum depth in which carbonate rocks can form is determined by the calcium carbonate compensation zone, generally beginning at a depth of about 2 miles. Below this zone, the cold, high-pressure waters of the abyss, which contain the vast majority of free carbon dioxide, dissolves calcium carbonate sinking to this level. The upwelling of deep ocean water, mainly in the tropics, returns to the atmosphere carbon dioxide lost by the carbon cycle, which is the circulation of carbon by geochemical processes.

Silica readily dissolves in seawater in volcanically active areas on the ocean floor, from volcanic eruptions into the sea, and from weathering of siliceous rocks on the continents. Some organisms such as sponges and diatoms extract the dissolved silica directly from seawater to build their shells and skeletons. Accumulations of siliceous sediment on the seafloor from dead organisms form diatomaceous earth also called diatomite. Thick deposits throughout the world are a tribute to the prodigious growth of these organisms over geologic time.

Seawater contains about 3.5 percent dissolved minerals, mostly sodium chloride, which precipitates into halite or common salt. Salt has been mined extensively throughout the world since ancient times. It is extracted from formerly shallow, stagnant pools of seawater called brines that have evaporated, which is why they are called evaporite deposits. The evaporation occurs in shallow, slowly sinking basins that are partially dammed by sandbars. During storms, seawater flows over the sandbars and replenishes the basins. Salt also accumulates in thick beds in deep basins that were cut off from the general circulation of the ocean.

As seawater evaporates, the concentration of salt increases to the saturation point. The salt then precipitates out of solution and accumulates on the seafloor. Several thousand feet of seawater evaporates to produce 100 feet of salt. However, many salt deposits are much thicker than this, possibly signifying many cycles of evaporation.

Several other minerals are also deposited, including gypsum, used in plaster and wallboard; phosphates and nitrates, used in fertilizers and explosives; potassium, used in many products such as fertilizers; and important halogens such as bromine, chlorine, and iodine. Evaporite deposits in the interiors of continents, such as the potassium deposits at Carlsbad, New Mexico, (Fig. 29) indicate these areas were once inundated by the sea. Many oil fields are located near ancient salt domes, which provide the geologic structures needed for trapping oil and gas.

Figure 29 *The Duval Sulfur and Potash Company's mining operation near Carlsbad, New Mexico.*

(Photo by E. F. Patterson, courtesy USGS)

Evaporite deposits generally form under arid conditions between 30 degrees north and 30 degrees south of the equator. Extensive salt deposits are not being formed presently, however, suggesting a relatively cooler global climate. That ancient evaporite deposits exist as far north as the Arctic regions indicates either these areas were once closer to the equator or the global climate was much warmer in the geologic past. Evaporite accumulation peaked about 230 million years ago, when the supercontinent Pangaea was in existence. Few evaporite deposits date beyond 800 million years ago, probably because most of the salt formed before then has been recycled.

Presently, the Mediterranean Sea is practically an enclosed basin. The evaporation rate is very high, and nearly 5 feet of the water's surface evaporate every year. This generates water with a high salt content, making it heavier than normal seawater. The highly saline water sinks to the bottom, and over time it will eventually fill the entire basin. Furthermore, the inflow from rivers into the Mediterranean cannot compensate for the evaporation and the outflow of water at the Gibraltar shelf.

The salts precipitate out of solution in stages. The first mineral to precipitate is calcite, closely followed by dolomite, although only minor amounts of limestone and dolostone are deposited in this manner. After about two-thirds of the water has evaporated, gypsum precipitates. When nine-tenths of the water is removed, halite forms. Thick deposits of halite are also produced by the direct precipitation of seawater in deep basins that have been cut off from the ocean.

Thick beds of gypsum, composed of hydrous calcium sulfate, constitute one of the most common sedimentary rocks. Gypsum is produced in evaporite deposits that formed when a pinched-off portion of the ocean or an inland sea evaporated. Oklahoma, as with many parts of the interior of North America invaded by a Mesozoic sea, is well known for its gypsum beds. The mineral is mined extensively for the manufacture of plaster, which was first used by the ancients in Asia for flooring material, containers, sculptures, and ornamental beads as early 12,000 B.C., long before the invention of pottery.

SEDIMENTARY STRUCTURES

Layers of sedimentary rock are separated by bedding planes, which are areas of weakness where the rocks tend to separate or break apart. The varying thicknesses of the layers reflect different depositional environments at the time the sediments were laid down. Each bedding plane generally marks where one type of deposit ends and another begins. Thus, thick sandstone beds might be interspersed with thin beds of shale, indicating that periods of coarse sediment deposition were punctuated by periods of fine sediment deposition possibly caused by changing climate conditions.

Graded bedding occurs when the particles in a sedimentary bed vary from coarse at the bottom to fine at the top. This type of bedding is indicative of rapid deposition of sediments of differing sizes by a fast-flowing stream emptying into the sea. The largest particles settle out first and are covered with progressively finer material due to the difference in settling rates. Beds also grade laterally, producing a gradation of sediments called a facies change from the Latin word for "form."

The color of sedimentary beds helps to identify the type of depositional environment. Generally, sediments tinted various shades of red and brown indicate a terrestrial source, whereas green and gray sediments suggest a marine environment. The size of individual particles also influences the color intensity, and generally darker-colored sediments indicate finer grains.

Figure 30 *Ripple marks on Dakota Sandstone, Jefferson County, Colorado.*
(Photo by J. R. Stacy, courtesy USGS)

Fluvial, or river, deposits are recognized in outcrops by their coarse sediment grains and cross-bedding features, generated when a stream meandered back and forth over old river channels. River currents also can align mineral grains and fossils, giving rocks a linear structure that can be used to determine the direction of current flow. Ripple marks on exposed surfaces (Fig. 30) are also used for determining the direction of river flow or wind direction if composed of desert sand. Sand dunes, when lithified, display a distinct dune structure on outcrops (Fig. 31), which can be used to determine the direction of travel.

Turbidities are sedimentary structures formed by turbidity currents. Submarine slides are responsible for carving out deep canyons in the seabed. The slides consist of sediment-laden water, and because it is heavier than the surrounding seawater it moves swiftly along the ocean floor and erodes the soft bottom material. These muddy waters, called turbidity currents, can move

Figure 31 *A trench at the crest of a 100-foot-high star dune southeast of Zalim, Saudi Arabia showing wedge-planar cross-strata dipping steeply in three directions.*

(Photo by E. D. McKee, courtesy of USGS)

down the gentlest slopes, transporting large rocks. Turbidity currents are also initiated by river discharge, coastal storms, or other currents. They play an important role in building up the continental slope and the smooth ocean bottom beyond.

After an assessment on erosion and sedimentation and their effects on the land, the next chapter will focus attention on the type sections that define the Earth's geologic features.

3

TYPE SECTIONS

DEFINING ROCK FORMATIONS

This chapter discusses the dating, correlation, and mapping of geologic formations. The Earth's many geologic layers, or stratigraphic units, have a complex system of classification. Stratigraphic units are classified into erathems, consisting of rocks formed during an era of geologic time. Erathems are divided into systems, consisting of rocks formed during a period of geologic time. Systems are divided into groups, consisting of rocks of two or more formations that contain common features. Formations are classified by distinctive features in the rock and are given the name of the locality where they were originally described. Formations are divided into members, which might be subdivided into individual beds such as sandstone, shale, or limestone.

A type section is a sequence of strata that was originally described as constituting a stratigraphic unit and serves as a standard of comparison for identifying similar widely separated units. Preferably a type section is selected in an area where both the top and bottom of the formation are exposed. Type sections are named for the area where they are best exposed. For example, the Jurassic Morrison Formation, well known for its dinosaur bones (Fig. 32), is named for the town of Morrison near Denver, Colorado.

Figure 32 *The Jurassic Morrison Formation in the Uinta Mountains, Utah.*

Type sections are also distinguished by their distinct fossil content used to correlate stratigraphic units. These are placed in order by age into a geologic column and are used to establish a geologic time scale. The use of dated material places actual ages on units of geologic time. Finally, a geologic map presents in plan the geologic history of an area where particular rock exposures are found.

GEOLOGIC TIME

The history of the Earth has been divided into units of geologic time according to the type and abundance of fossils present in the strata. The major units were delineated by 19th-century geologists in Great Britain and western Europe. The periods take their names from the localities with the best exposures. For example, the Jurassic period was named for the Jura Mountains in Switzerland, whose limestones provide a suite of fossils that adequately depicts the period.

Since geologists had no means of determining the actual ages of rocks, they created the entire geologic record using relative dating techniques, which places geologic time units in their proper sequence without reference to their actual age. Since then, absolute dates have been added to units of geologic time following the development of radiometric dating techniques based on the decay of radioactive isotopes.

Geologic time is divided into eras. They include the Precambrian, known as the age of prelife; the Paleozoic, known as the age of ancient life; the Mesozoic, known as the age of middle life; and the Cenozoic, known as the age of new life. The longest era, the Precambrian, is 4 billion years in duration and mostly obscure owing to the scarcity of fossilized remains of ancient organisms. At the beginning of the Paleozoic era, about 570 million years ago, the fossil record vastly improved due to the proliferation of species with hard skeletons. Before this time, organisms were soft-bodied and therefore did not fossilize well.

The lack of fossils in ancient rocks often puzzled early geologists. Then suddenly life appeared in rocks at the base of the Cambrian period in great abundance at the same horizon the world over (Fig. 33). The Cambrian period was named for the Cambrian Mountains in central Wales, where sediments

Figure 33 *A syncline in weathered shale of the lower Cambrian Rome Formation, Johnson County, Tennessee.*

(Photo by W. B. Hamilton, courtesy USGS)

containing the earliest known fossils were found. The base of the Cambrian was thought to be the beginning of life, and all time before then was simply called Precambrian.

The eras following the Precambrian are divided into smaller units called periods. The Paleozoic has seven periods, the Mesozoic has three, and the Cenozoic has two. Each period is characterized by changes in organisms that are less profound compared with the eras, which mark the boundaries of mass extinctions, proliferations, or rapid transformations of species. The two periods of the Cenozoic are further subdivided into seven epochs because of the greater detail provided by recent rocks. The epoch we live in is called the Holocene, which corresponds to the Neolithic in archaeology and the beginning of civilization.

AGE OF THE EARTH

The age of the Earth can be compared with the length of a single day. Half an hour after midnight, the Earth emerged from a collection of primordial dust and gas. Life first appeared about 3:00 A.M. By 4:00 P.M., the first single-celled animals evolved. Multicellular animals called metazoans arrived around 8:00 P.M. The first vertebrates followed one hour later and conquered the land less than an hour after that. The dinosaurs turned up about 11:00 P.M. They were supplanted by the mammals about a half hour later. One minute to midnight humans arrived.

The sixth-century B.C. Greek philosopher Xenophanes was perhaps the first to speculate on the age of the Earth. He believed that rocks containing fossil seashells in the mountainsides were evidence that they must have originated in the sea below. He thought the Earth had to be extremely old to allow enough time for the imperceptible growth of mountains. Otherwise, some spectacular catastrophe would have occurred that rapidly raised them high above the sea.

Early geologists tried many methods to determine the age of the Earth. One was to measure the thickness of sedimentary strata and compare it to known sedimentation rates. This method was notoriously unreliable and provided varying dates, depending on the locality because some deposits were much thicker than others. Another method was to compare the salinity of rivers with that of the ocean, which was thought to have started fresh and became salty over time. These techniques gave the Earth an age of roughly 100 million years, a date that was generally accepted by most 19th-century geologists.

Another approach was to calculate the time required for the Earth to form out of the solar nebula and cool from a molten state to its present temperature. However, this method was soon discarded after the discovery of

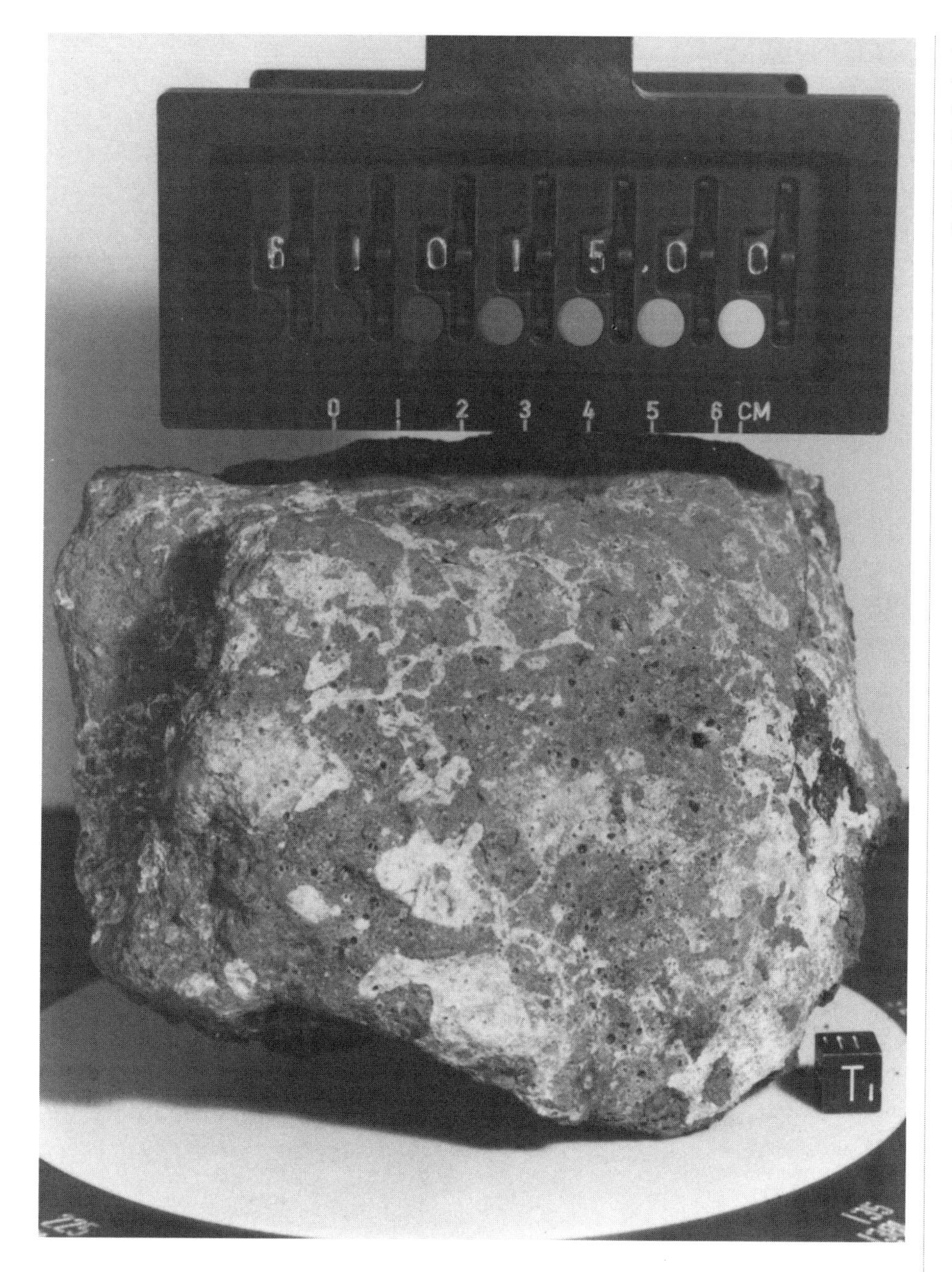

Figure 34 *A lunar sample brought back by Apollo astronauts, similar to ejected fragments around impact craters.*

(Photo by H. J. More, courtesy USGS)

radioactive isotopes, which generated heat as they decayed into stable elements. This heat source was responsible for maintaining the Earth's interior temperature since the very beginning. By calculating the half-lives of certain radioactive isotopes found on the Earth, meteorites, and moon rocks (Fig. 34), scientists determined that the Earth is about 4.6 billion years old.

FAUNAL SUCCESSION

The history of the Earth is written in the rocks, and the history of life is told by fossils. However, the fossil record is not complete because the remaking of the surface has erased entire chapters of geologic time. Yet the study of fossils along with radiometrically dating the rocks that contain them has enabled geologists to construct a reasonably good chronology of Earth history. The fossil record also provides valuable insights into the evolution of the Earth. Moreover, knowledge of the origination and extinction of species throughout the fossil record is important for building an accurate account of the evolution of species through time (Table 4).

Although the existence of fossils has been known since the ancient Greeks first discovered seashells in the hillsides far from shore and pondered how they got there, the discovery of their significance as a geologic tool was not made until the late 18th century. In the 1790s, the English civil engineer William Smith found that rock formations in the canals he excavated across Britain contained fossils that were different from those in beds above or below them. He also noticed that sedimentary strata in widely separated areas could be identified by their distinctive fossil content. Smith drew geologic maps of the varied rock formations throughout Britain by using the characteristics of the different strata and their fossils. He put forward the idea that two rock layers from different sites could be regarded as equivalent in age as long as they contained the same fossils. Furthermore, one type of bed such as sandstone

TABLE 4 EVOLUTION OF THE BIOSPHERE

	Billions of Years Ago	Percent Oxygen	Biologic Effects	Event Results
Full oxygen conditions	0.4	100	Fishes, land plants and animals	Approach present biologic environs
Appearance of shelly animals	0.6	10	Cambrian fauna	Burrowing habitats
Metazoans appear	0.7	7	Ediacaran fauna	First Metazoan fossils & tracks
Eukaryotic cells appear	1.4	>1	Larger cells with a nucleus	Redbeds, multicellular organisms
Blue-green algae	2.0	1	Algal filaments	Oxygen metabolism
Algal precursors	2.8	<1	Stromatolite mounds	Initial photosynthesis
Origin of life	4.0	0	Light carbon	Evolution of the biosphere

might grade into a different bed such as limestone that contains identical fossils, indicating they were the same age. These observations led him to propose the law of faunal succession, one of the most important and basic principles of historical geology.

Around the same time, the French geologists Georges Cuvier and Alexandre Brongniart found that certain fossils in the rocks around Paris were restricted to specific beds. The geologists arranged their fossils in chronologic order and discovered that they varied in a systematic way according to their positions in the strata. Fossils in the higher rock layers more closely resembled modern forms of life than those further down in the geologic column. Also, the fossils did not occur randomly but in a determinable order from simple to complex. Therefore, geologic time periods could be identified by their distinctive fossil content. This became the basis for establishing the geologic time scale (Table 5) and the beginning of modern geology.

In 1830, the British geologist Charles Lyell took these ideas one step further, by proposing that rock formations and other geologic features took shape, eroded, and reformed at a constant rate throughout time according to the theory of uniformitarianism, which states that the present is the key to the past. In other words, the forces that shaped the Earth are uniform and operated in the past much as they do today. The theory was originally developed in 1785 by Lyell's mentor, the Scottish geologist James Hutton, known today as the "father of geology." Hutton envisioned the prime mover behind these slow changes to be the Earth's own internal heat. Geologists had long recognized that rocks were molten in the Earth's interior, and this observation was manifested by volcanic eruptions.

Hutton's discovery of unconformities, places where ancient sedimentary strata were upturned, eroded, and blanketed by younger deposits, suggested the history of the Earth is exceedingly long and complex. Hutton believed contemporary rocks at the surface were formed by the waste of older rocks laid down in the sea, consolidated under great pressure, and upheaved by the expanding power of the Earth's subterranean heat. His theory was based on the premise that the depths of the Earth are in a constant state of turmoil and that molten matter rises to the surface through cracks or fissures, resulting in an erupting volcano.

Lyell continued with Hutton's work and gained worldwide acceptance for the theory of uniformitarianism. He gathered many observations about rocks and landforms of western Europe, showing they were the products of the same processes in existence today provided they were given enough time. Many geologists, however, felt this theory was not fully adequate to explain all the geologic forces at work. Events in the past appeared not to be slowly evolving but occurred suddenly according to the fossil record, which showed at times catastrophic extinctions of species.

TABLE 5 THE GEOLOGIC TIME SCALE

Era	Period	Epoch	Age (Millions of Years)	First Life Forms	Geology
		Holocene	0.01		
	Quaternary				
		Pleistocene	3	Humans	Ice age
Cenozoic		Pliocene	11	Mastodons	Cascades
		Neogene			
		Miocene	26	Saber-toothed tigers	Alps
	Tertiary	Oligocene	37		
		Paleogene			
		Eocene	54	Whales	
		Paleocene	65	Horses, Alligators	Rockies
	Cretaceous		135		
				Birds	Sierra Nevada
Mesozoic	Jurassic		210	Mammals	Atlantic
				Dinosaurs	
	Triassic		250		
	Permian		280	Reptiles	Appalachians
	Pennsylvanian		310		Ice age
				Trees	
	Carboniferous				
Paleozoic	Mississippian		345	Amphibians	Pangaea
				Insects	
	Devonian		400	Sharks	
	Silurian		435	Land plants	Laursia
	Ordovician		500	Fish	
	Cambrian		570	Sea plants	Gondwana
				Shelled animals	
			700	Invertebrates	
Proterozoic			2500	Metazoans	
			3500	Earliest life	
Archean			4000		Oldest rocks
			4600		Meteorites

When fossils are arranged according to their age, they do not present a random or haphazard picture, but instead show progressive changes from simple to complex life forms and reveal the advancement of species through time. Geologists were thus able to recognize geologic time periods based on groups of organisms that were especially plentiful and characteristic during a particular time. Within each period, many subdivisions were determined by the occurrence of certain species, and this succession is found on every major continent and is never out of order.

The branch of geology devoted to the study of ancient life based on fossils is called paleontology. Fossils are the remains or traces of organisms preserved from the geologic past. Not all organisms become fossils, however, and plants and animals must be buried under stringent conditions to become fossilized. Given enough time and the proper conditions, the remains of an organism are modified, often becoming petrified and literally turned to stone (Fig. 35).

Fossils are important for correlating rock units over vast distances. Since certain species have lived only during specific times, their respective fossils can be used to place stratigraphic units in their proper sequence or relative time periods. These beds can then be traced over wide areas by comparing their fossil content. This provides a comprehensive geologic history over a broad region and establishes a relative time scale that can be applied to all parts of the world. Absolute dates have since been placed on units of relative time to further improve the understanding of geologic history.

Figure 35 *Fossilized tree trunks at Petrified Forest National Park, Apache County, Arizona.*

(Photo by N. H. Darton, courtesy USGS)

RELATIVE TIME

A major problem encountered when exploring for fossils of early life is that the Earth's crust is constantly rearranging itself, and only a few fossil-bearing formations have survived undisturbed over time. The history of the Earth as told by its fossil record is therefore incomplete because the remaking of the surface has erased whole chapters of geologic history.

The existence of fossils has been known since ancient times. The Greek philosopher Aristotle clearly recognized that certain fossils were the remains of organisms, although he generally believed that fossils were the result of some celestial influence. This astrological account for fossils remained popular throughout the Middle Ages. During the late Renaissance period, which ushered in a rebirth of science, alternative explanations for the existence of fossils based on scientific principles were sought. By the 18th century, most scientists began to accept fossils as the remains of organisms because they more closely resembled living things rather than merely inorganic substances.

Nineteenth-century geologists used fossils to define the boundaries of the geologic time scale. However, because they had no means of actually dating the rocks that contained the fossils, geologists delineated the entire geologic record using units of relative time. This dating method only indicated which bed was older or younger according to its fossil content. Therefore, relative dating placed rocks in their proper sequence or order but did not indicate how long ago an event took place, only that it followed one period and preceded another.

Geologists measure geologic time by tracing fossils through the rock strata and observing a greater change with deeper rocks as compared with those near the surface. Fossil-bearing strata can be followed horizontally over great distances because a particular fossil bed can be identified in another locality with respect to the beds above and below it. These marker beds are used for identifying geologic formations and to delineate rock units for geologic mapping.

One of the most frustrating aspects of dealing with the geologic time scale are gaps in the fossil record, where portions of Earth history have been erased. This break in geologic time might have resulted from periods of erosion or nondeposition of sedimentary strata that traps and preserves species as fossils. Gaps in the fossil record also might be attributed to insufficient intermediary species, or so-called missing links, which might have existed only in small populations. Small populations are less likely to leave a fossil record, and usually the process of fossilization favors large populations of species, which are given the greatest recognition in the fossil record.

ROCK CORRELATION

The 17th-century Danish physician and geologist Nicholas Steno recognized that in a sequence of rock layers undeformed by folding or faulting each bed was formed after the one below it and before the one above it. This became known as the law of superposition. Steno also put forward the principle of original horizontality, which states that sedimentary rocks were originally laid down in the ocean horizontally, and subsequent folding and faulting uplifted them out of the sea and inclined them at steep angles.

If angled rocks are overlain by horizontal ones, they represent a gap in time known as an angular unconformity (Fig. 36). Furthermore, if a body of rocks cuts across the boundaries of other rock units, it is younger than those it intercepts. This is the principle of cross-cutting relationship, which states that granitic intrusions are younger than the rocks they invade. A sequence of rocks placed in their proper order is called a stratigraphic cross section (Fig. 37).

To develop a geologic time scale applicable over the entire world, rocks of one locality are matched or correlated with rocks of similar age in another

Figure 36 *An angular unconformity between the Chacarilla and the Altos de Pica formations, Andes Mountains, Tarapaca Province, Chile.*

(Photo by R. J. Dingman, courtesy USGS)

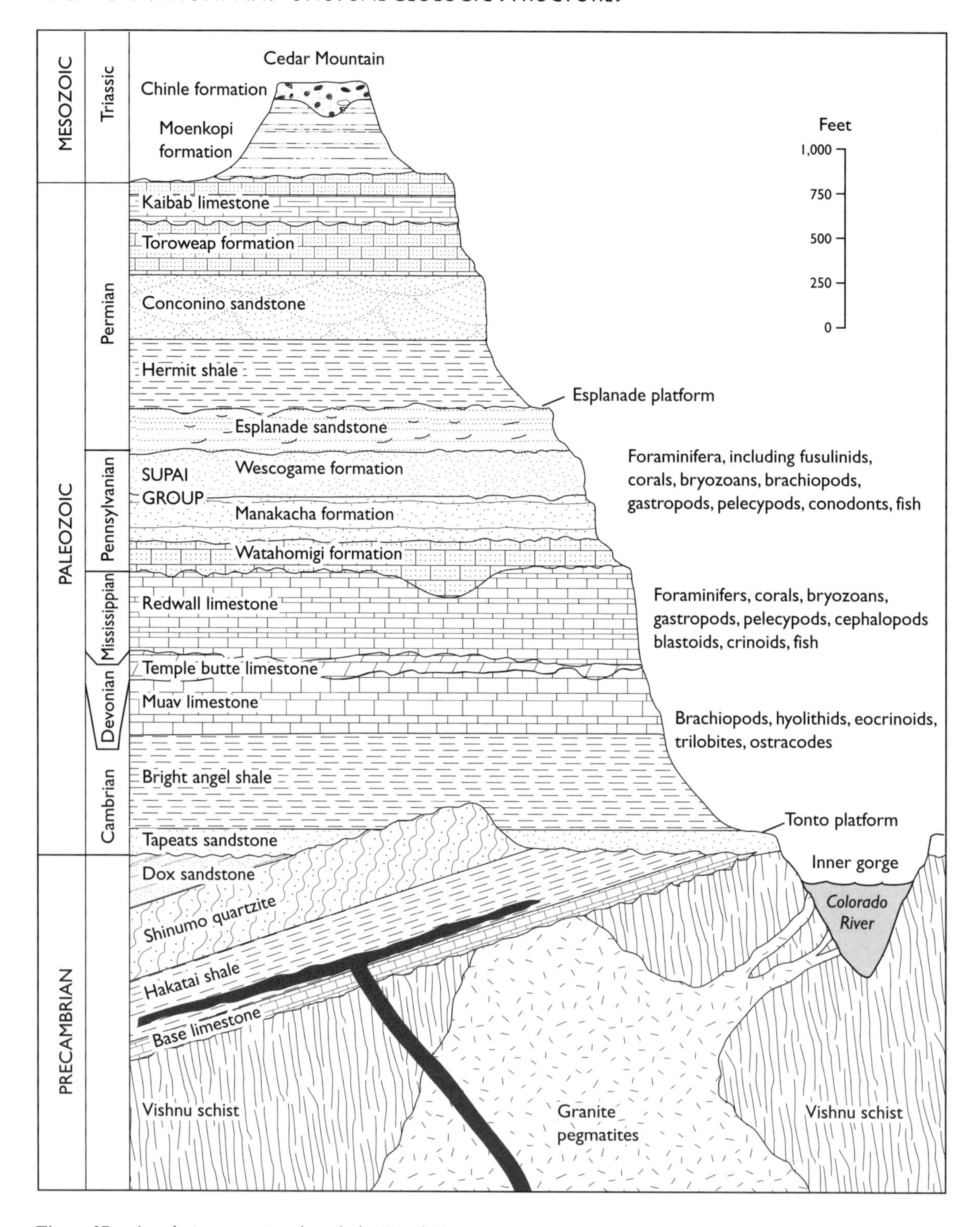

Figure 37 *A geologic cross section through the Grand Canyon.*

locality. By correlating rocks from one place to another over a wide area, geologists can obtain a comprehensive view of the geologic history of a region. A bed or a series of beds can thus be traced from one outcrop to another by recognizing certain distinctive features in the rocks.

A problem arises, however, if two or more identical rock units exist at each locality, causing difficulties in matching beds. To complicate matters further, if faulting occurred in the area, one block of a rock sequence might be down-dropped in relation to the other or thrust over another. A stratum folded over on itself contains rock units that are completely reversed (Fig. 38), making matters even more confusing. Rocks occurring in repetitive sequences of sandstone, shale, and limestone complicate correlation even further.

Although these methods might be sufficient to trace rock formations over relatively short distances, they are inadequate for matching rocks over long distances, such as from one continent to another. Therefore, to correlate between widely separated areas or between continents, fossils were used. Later, geologic dates were added to the rocks that contained them to further refine rock correlation.

DATING ROCKS

Before the advent of radiometric dating, geologists could not date geologic events precisely. Therefore, relative dating techniques that relied on the fossil content of the rocks were developed and are still in use today. Absolute dating methods did not replace these techniques, but only supplemented them. Since accurate absolute dates have been applied to periods of relative time, howev-

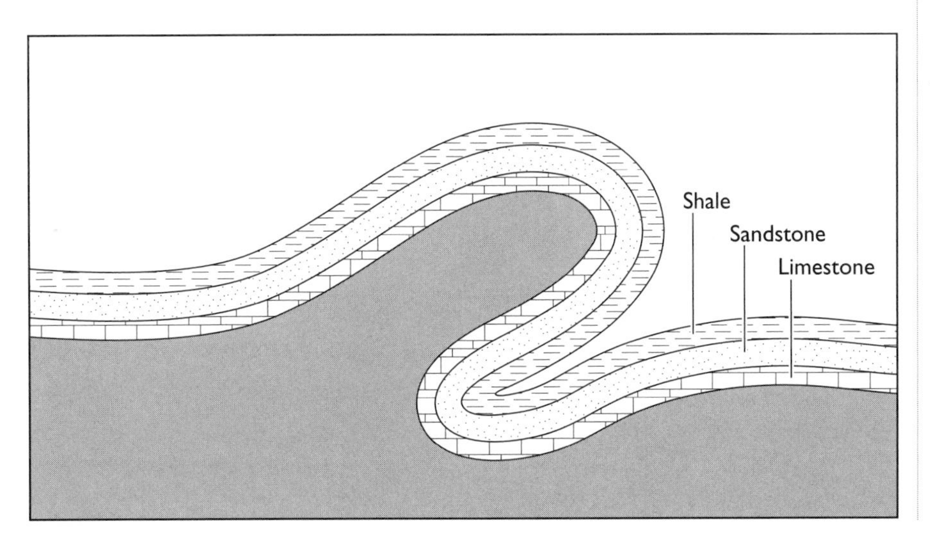

Figure 38 *Intensely folded strata can make correlation difficult.*

er, some difficulties have been encountered, including disagreements over the dates of certain events in geologic history.

The basic problem in attempting to assign absolute dates to units of relative time is that most radioactive isotopes are restricted to igneous rocks. Even if sedimentary rocks, which comprise most of the rocks on the Earth's surface and contain all the fossils, did contain a radioactive mineral, most could not be dated accurately because the sediments were composed of grains derived from older rocks. Therefore, sedimentary rocks generally cannot be dated directly by radioactive decay. To date sedimentary rocks they have to be related to igneous masses. Volcanic ash deposited in a layer above or below a sediment bed could be dated radiometrically, as well as cross-cutting features such as a granitic dike, which is younger than the beds it crosses. The sedimentary strata would then be bracketed by dated materials, and its age could be determined fairly accurately.

Radiometric dating methods measure the ratios of radioactive parent materials to their daughter products and compare this ratio with the known half-lives of the radioactive elements. The half-life is the time required for one-half of a radioactive element to decay to a stable daughter product. For example, if one pound of a hypothetical radioactive element had a half-life of 1 million years, then after a period of 1 million years a half pound of the original parent material and a half pound of daughter product would be present.

The ratio of parent element to its daughter product is determined by chemical and radiometric analysis of the sample rock. Therefore, if the quantities of parent and daughter are equal, one half-life has expired, making the sample 1 million years old. After 2 million years, one-quarter of the original parent element remains in the sample, and after 4 million years only one-sixteenth is left. Generally, radioactive elements are usable for age dating up to 10 half-lives. Afterward, the amount of parent material is reduced to about a thousandth of its original mass.

Radioactive decay also appears to be constant with time and is unaffected by chemical reactions, temperature, pressure, or any other known conditions or processes that could change the decay rate throughout geologic history. Confirmation that decay rates are steady throughout time is found in certain minerals such as biotite mica. Extremely small zones of discoloration, or haloes, are found surrounding minute inclusions of radioactive particles within the crystal. The haloes consist of a series of concentric rings around the radioactive source. Particles emitted by the radioactive source damage the surrounding biotite minerals. The energy of the particle is determined by how far it travels through the mineral and depends on the type of radioactive element responsible. Since the radii of concentric rings correspond to the energy levels of present-day particles, particle energies do not appear to have changed and the rate of radioactive decay remains constant with time.

The precision of radiometric age dating depends on the accuracy of the chemical analysis that determines the amount of the parent element and daughter product, and whether either has been added to or removed from the sample since deposition. The quantities of these substances might only be a few parts per million of the rock mass. A certain amount of naturally occurring daughter material might have existed in the rock before the parent element began decaying. Moreover, many radioactive elements do not decay directly into stable daughter products, but go through a series of intermediate decay schemes, further complicating analysis.

To date recent events, scientists use a radioactive isotope of carbon called carbon 14, or radiocarbon. Carbon 14 is continuously created in the upper atmosphere by cosmic ray bombardment of gases, which in turn release neutrons. The neutrons bombard nitrogen in the air, causing the nucleus to emit a proton, thus converting nitrogen to radioactive carbon 14. In chemical reactions, this isotope behaves identically to natural carbon 12. It reacts with oxygen to form carbon dioxide, circulates in the atmosphere, and is absorbed directly or indirectly by living matter (Fig. 39). As a result, all organisms contain a small amount of carbon 14 in their bodies.

Carbon 14 decays at a steady rate with a half-life of 5,730 years. While an organism is alive, the decaying radiocarbon is continuously being replaced, and the proportions of carbon 14 and carbon 12 remain constant. However, when a plant or animal dies, it ceases to intake carbon, and the amount of carbon 14 gradually decreases as it decays to stable nitrogen 14. This results from the emission of a beta particle (free electron) from the carbon 14 nucleus, thus transmuting a neutron into a proton and restoring the nitrogen atom to its original state.

Radiocarbon dates are determined by chemical and radiometric analysis, which compares the proportions of carbon 14 to carbon 12 in a sample. The development of improved analytical techniques has increased the usefulness of radiocarbon dating, and it can be used to date events taking place more than 100,000 years ago. This makes it a valuable tool for dating events that occurred during the last ice age. Furthermore, paleontologists, anthropologists, archaeologists, and historians have a means of accurately dating events from humankind's distant past.

Of all the radioactive isotopes existing in nature, only a few have been proven useful in dating ancient rocks (Table 6). The others either are very rare or have half-lives that are too short or too long. Rubidium 87 with a half-life of 47 billion years, uranium 238 with a half-life of 4.5 billion years, and uranium 235 with a half-life of 0.7 billion years are useful for dating rocks that are tens of millions to billions of years old. The uranium isotopes are important for dating igneous and metamorphic rocks. Also, because both species of uranium occur together, they can be used to cross-check each other.

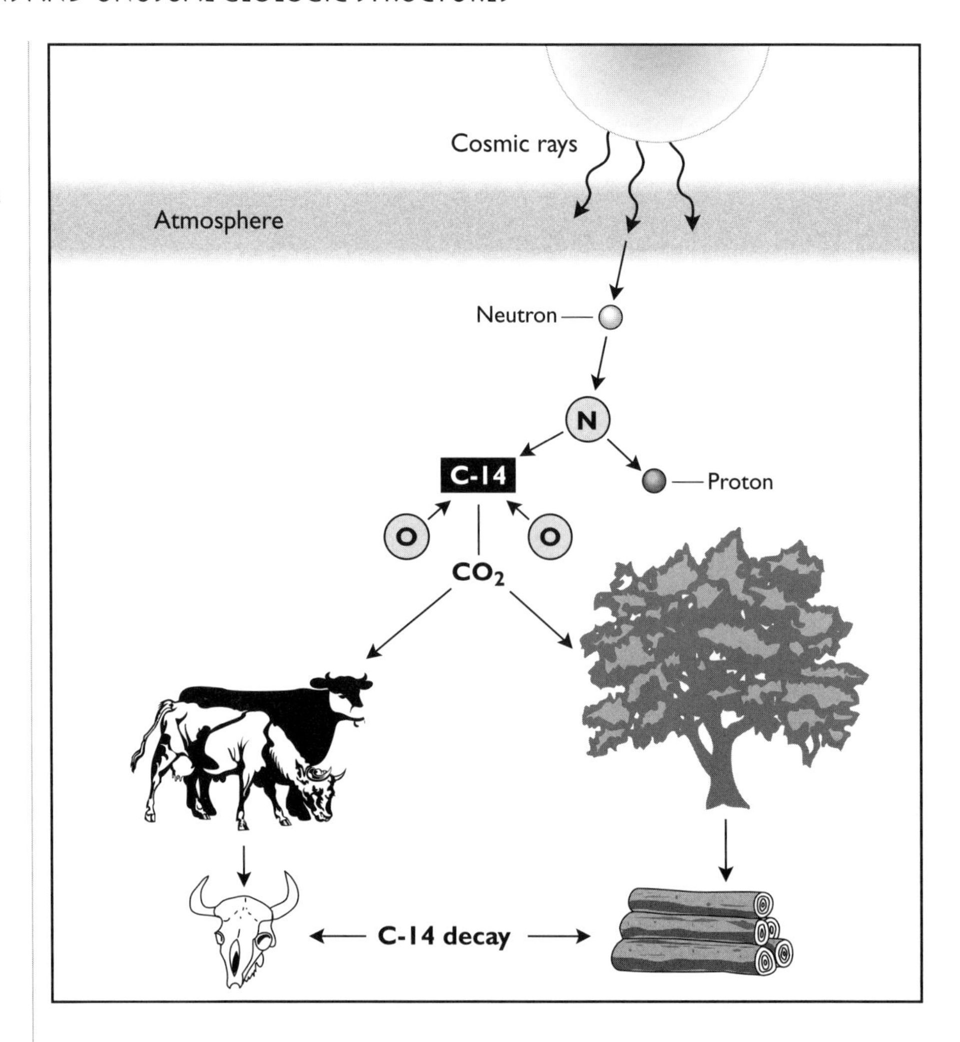

Figure 39 *The carbon 14 cycle. Cosmic rays striking the atmosphere releases neutrons that strike nitrogen atoms to produce carbon 14, which is converted into carbon dioxide and taken in by plants and animals.*

Zircon crystals found in granitic and volcanic rocks are enormously resistant to erosion and tell of the earliest history of the Earth, when the crust first formed some 4.2 billion years ago. A rock containing zircon crystals that hold uranium and its decay products can partially melt in a volcanic eruption millions of years after formation. The zircon can survive the melting and then grow a new layer of crystal over the old core. Therefore, when the whole crystal is analyzed, the apparent age will be older than the age of the volcanic ash bed where the zircon crystals are found, thereby confusing the dating method using the theory of cross-cutting relationships.

Potassium 40 is more versatile for dating younger rocks. Although the half-life of potassium 40 is 1.3 billion years, recent analytical techniques enable the detection of minute amounts of its stable daughter product argon 40 in

TABLE 6 FREQUENTLY USED RADIOISOTOPES FOR GEOLOGIC DATING

Radioactive Parent	Half-life (Years)	Daughter Product	Rocks and Minerals Commonly Dated
Uranium 238	4.5 billion	Lead-208	Zircon, uraninite, pitchblende
Uranium 235	713 million	Lead-207	Zircon, uraninite, pitchblende
Potassium 40	1.3 billion	Argon-40	Muscovite, biotite, hornblende, glauconite, sanidine, volcanic rock
Rubidium 87	47 billion	Strontium-87	Muscovite, biotite, lepidolite, microcline, glauconite, metamorphic rock
Carbon 14	5,730 million	Nitrogen-14	All plant and animal material

rocks as young as 30,000 years. It is less precise for dating younger rocks because of the small amount of daughter product available in the sample. Minerals such as hornblende, nepheline, biotite, and muscovite are used for dating most igneous and metamorphic rocks by the potassium-argon method.

Dating sedimentary rocks presents a more difficult problem because their material was derived from weathering processes. Fortunately, a micalike mineral called glauconite forms in the sedimentary environment and contains both potassium 40 and rubidium 87. As a result, the age of the sedimentary deposit can be established directly by determining the age of the glauconite. Unfortunately, metamorphism, no matter how slight, might reset the radiometric clock by moving the parent and daughter products elsewhere in the sample. Therefore, the radiometric measurement can date only the metamorphic event. To date these rocks accurately, a whole-rock analysis must be made using large chunks of rock instead of individual crystals. Sediments also can be dated using optically stimulated thermoluminescence, which measures when sand grains were last exposed to light, and is especially useful for dating fossil footprints.

GEOLOGIC FORMATIONS

Beginning at the very bottom of the geologic column are Precambrian granitic and metamorphic rocks (Fig. 40). These are overlain by progressively younger sediments, igneous intrusives, metamorphics, and volcanic extrusives. Deposited on top of the Precambrian basement rock are Proterozoic conglomerates, which are consolidated equivalents of sand and gravel. Nearly

Figure 40 *Sawatch sandstone resting on Precambrian granite on Ute Pass, El Paso County, Colorado.*

(Photo by N. B. Darton, courtesy USGS)

20,000 feet of Proterozoic sediments are found in the Uinta range of Utah (Fig. 41), which is the only major east-west trending mountain range in the lower United States. The Montana Proterozoic belt system contains sediments up to several miles thick.

The Proterozoic is also known for its terrestrial redbeds, so named because the sediment grains were cemented with iron oxide, which stained the rocks red. In the western United States, a preponderance of red rocks is exposed in the mountains and canyons. These sedimentary rocks were cemented by an iron oxide mineral called hematite, so named because of its blood red color. Redbeds are clear evidence that the Earth's atmosphere contained significant amounts of oxygen, which oxidized the iron in a way similar to rusting.

During the middle Paleozoic, as the continents rose higher and sea levels dropped lower, the inland seas departed and were replaced by great swamps. In these regions, thick coal deposits accumulated during the Carboniferous, which includes the Mississippian and Pennsylvanian periods in North America. The Carboniferous and the Permian periods had the highest organic burial rates in Earth history. A major ice age also occurred during the late Carboniferous.

The Permian witnessed the complete retreat of marine waters from the land, an abundance of terrestrial redbeds, and large deposits of gypsum and salt. In North America, extensive terrestrial redbeds covered the Colorado Plateau (Fig. 42) and a region from Nova Scotia to South Carolina. Redbeds were also common in Europe. The wide occurrence of red sediments might have resulted from massive amounts of iron supplied by intense igneous activity the world over. Air trapped in ancient tree sap suggests a greater abundance of atmospheric oxygen, responsible for oxidizing the iron into hematite.

Important reserves of phosphate used for fertilizers were laid down in the late Permian in Idaho and adjacent states. Huge sedimentary deposits of

Figure 41 *Lodore Canyon, looking north toward Browns Park, Uinta Mountains, Summit County, Utah.*

(Photo by W. R. Hansen, courtesy USGS)

iron were also laid down, but they were not nearly as rich as those of the Proterozoic. The ore-bearing rocks of the Clinton iron formation, the chief iron producer in the Appalachian region from Alabama to New York, were deposited during this time.

An inland sea, called the Western Interior Cretaceous Seaway, flowed into the west-central portions of North America. Accumulations of marine sediments eroded from the Cordilleran highlands to the west were deposited on the terrestrial redbeds of the Colorado Plateau, forming the Jurassic Morrison Formation, well known for its abundant dinosaur bones. Rabbit Valley, west of Grand Junction, Colorado, is a dinosaur quarry in Morrison rocks with easy access lying just off the interstate highway, where for more than 100 years fossil collectors have been striking it rich. The quarry is responsible for the discovery of one of the largest dinosaur genus called *Apathosaurus.* This gargantuan creature fully deserves the title "thunder lizard."

During the Cretaceous period, huge deposits of limestone and chalk were laid down in Europe and Asia. Seas invaded Asia, South America, Africa, Australia, and the interior of North America. Into these vast bodies of water

Figure 42 *Chugwater redbeds, Big Horn County near Shell, Wyoming.*

(Photo by G. A. Fisher, courtesy USGS)

Figure 43 *Columbia River basalt, looking downstream from Palouse Falls, Franklin-Whitman Counties, Washington.*

(Photo by F. O. Jones, courtesy USGS)

were deposited thick layers of sediment, which are presently exposed as impressive sandstone cliffs in the western United States and elsewhere.

During the Tertiary period, volcanic activity was extensive and great outpourings of basalt covered Washington, Oregon, and Idaho, creating the Columbia River Plateau (Fig. 43). Massive floods of lava poured onto an area of about 200,000 square miles and in places reached 10,000 feet thick. In only a matter of days, volcanic eruptions spewed out batches of basalt as large as 1,200 cubic miles, forming lava lakes up to 450 miles across. Massive volcanism also occurred in other parts of the world, prompting the volcanic theory of dinosaur extinction, because such an environmental catastrophe would have made living conditions intolerable for many species.

GEOLOGIC MAPPING

To appreciate the geologic history of an area, a basic understanding of geologic maps is needed. Many geologic forms and structures are associated with a particular rock type and can often be recognized some distance away. Therefore, the size, shape, and composition of landforms depend on the nature of the rocks that comprise them. Rocks that form mountains or underlie valleys become visible when pushed or folded upward or when they jut through the ground and become easily accessible outcrops.

Landforms are the basic features of the Earth and include cliffs, hills, canyons, valleys, plateaus, and basins. They often express the type of rock that comprises them. All landforms are the result of a combination of processes that continually build up the land surface and tear it down. The knowledge of landforms and structures is therefore fundamentally important for interpreting the landscape and the geologic history of a region.

Geologic maps display the distribution of rocks on the Earth's surface. They also indicate the relative ages of these formations and profile their position beneath the surface. Often, much of the information is compiled from just a few available exposures, which must be extrapolated over a large area. The first geologic maps were made by geologists in Britain, where one of their first practical uses was for the exploration of coal. In the western United States, early explorers were amazed by the magnificent rock exposures. Pioneer geologists like John Wesley Powell (Fig. 44), who was the first to explore the Grand Canyon, made extensive geologic maps of the West, often sketching the formations as they rode through the region on horseback.

Modern geologic maps incorporate field observations and laboratory measurements, which are limited by rock exposures, accessibility, and personnel. Regional geologic maps present rock composition, structure, and geologic age, which are essential for constructing the geologic history of an area. Aided by geophysical data used for defining subsurface structures (Figs. 45a, 45b), these reconstructions are important because formations of rock units and geologic structures influence the deposition of much of the mineral wealth of the world.

Remote sensing methods are primarily used for augmenting conventional techniques of compiling and interpreting geologic maps of large regions. Remote sensing techniques enable geologists to obtain certain structural and lithologic information much more efficiently than can be achieved on the ground. In well-exposed areas, geologic maps can be made from aircraft and satellite imagery, even when only limited field data are available because many major structural and lithologic units are well displayed on the imagery.

Lineaments, long, linear trends in the Earth's surface, are one of the most obvious and most useful features in the imagery. Lineaments represent zones of weakness in the Earth's crust, often a result of faulting. Lineaments and texture along with dip and strike of the strata, which are the degree and direction of formation slope, further aid in geologic mapping.

Other features frequently observed in the imagery include circular structures created by domes, folds, and intrusions of igneous bodies into the crust.

Figure 44 *Major John Wesley Powell.*

(Photo courtesy USGS)

Structural features such as folds, faults, dips, and strikes of particular rock formations, along with lineaments, landform features, drainage patterns, and other anomalies might suggest areas where oil traps exist.

Stream drainage patterns, which are influenced by topographic relief and rock type, give additional clues about the type of geologic structure. Furthermore, the color and texture of the structure carry information about the rock formations it comprises. With this information, geologists can gen-

Figures 45a, 45b
Vibrosis vehicle and seismic instrument truck are used for defining subsurface structures.

(Photos courtesy U.S. Department of Energy)

erate large-scale geologic maps of inaccessible areas. This development has the potential of opening entirely new areas for mineral and petroleum exploration, especially in an age when the need for resources is becoming highly critical.

After defining rock formations, the next chapter will examine the folding and faulting of the Earth's rocks.

4

FOLDING AND FAULTING

THE SHAPING OF THE LAND

This chapter examines the forces that shape the Earth's crust. Tectonics, from the Greek *tekton* meaning "builder," is responsible for the Earth's active geology, including tall mountains, long rift valleys, and deep-ocean trenches. Plate tectonics creates the most impressive geologic features on the planet, using the forces of uplift combined with erosion. The crust is also sliced up by faults, which relieve pressures building up from plate motions. Slippage along major fault systems is often accompanied by earthquakes that can be very destructive as they wrench the landscape apart. Indeed, many places in the world are on shaky ground.

When a continent is stretched by tectonic forces, originating deep within the Earth, massive chunks of crust bounded by faults collapse, creating downdropped fault blocks (Fig. 46). This further weakens the crust and sets the stage for additional faulting. These downdropped fault blocks are often associated with upraised sections of crust, providing a landscape of ridges and troughs. Not all faults are vertical, however, and many result from horizontal forces produced by the compression of the crust or by plates shearing past each other. The movement of the crust along faults causes earthquakes, which rupture the crust and rearrange the landscape.

Figure 46 *Erratic course of the North Fork of the Payette River at the south end of the Cascade fault-block basin, Valley County, Idaho.*

(Photo by D. L. Schmidt, courtesy USGS)

TECTONISM

The Earth's outer shell is fashioned out of eight major and several minor movable plates that account for the tectonic activity taking place on the surface of the planet. The plates comprise the lithosphere, the rigid outer layer of the mantle, and the overlying continental or oceanic crust. Because continental crust is made of light materials, it remains on the surface, where it continues to grow by collecting crustal fragments. The oceanic crust, on the other hand, because it is made of denser materials, subducts deep into the mantle at ocean trenches, where it remelts in a continuous cycle.

The plate boundaries are the spreading ridges where new oceanic crust is created, the deep-sea trenches where old oceanic crust is subducted into the mantle and destroyed, and the transform faults where plates slide past each other. The plates carry the continents along with them as they ride on the semimolten rocks of the asthenosphere, the pliable layer of the upper mantle beneath the lithosphere. When two plates collide, they create mountain ranges on the continents and volcanic island arcs on the ocean floor. When an oceanic plate subducts beneath a continental plate, it forms sinuous mountain chains, such as the Andes of South America, and volcanic mountain ranges, such as the Cascades of the Pacific Northwest, noted for their powerful eruptions (Fig. 47). The breakup of a plate creates new continents and oceans. This process of rifting and patching of continents has been ongoing for at least 2.7 billion years.

An impressive submarine mountain range known as the Mid-Atlantic Ridge runs along the middle of the Atlantic Ocean, surpassing in scale the Alps and Himalayas combined. It is part of a global spreading-ridge system that stretches more than 40,000 miles along the ocean floor like the stitching on a baseball. A deep trough, resembling a giant crack in the Earth's crust, is carved down the middle of the ridge. It is the longest and deepest canyon on Earth.

The Mid-Atlantic Ridge is also the center of intense seismic and volcanic activity and the focus of high heat flow from the Earth's interior. Molten magma originating from the mantle rises through the lithosphere and erupts on the ocean floor, adding new oceanic crust to both sides of the ridge crest. Meanwhile, the upwelling magma pushes apart the two lithospheric plates, upon which ride the continents that surround the Atlantic Ocean.

Figure 47 *Eruption cloud from Mount St. Helens during the height of major eruptive activity on May 18, 1980.*

(Photo by A. Post, courtesy National Forest Service)

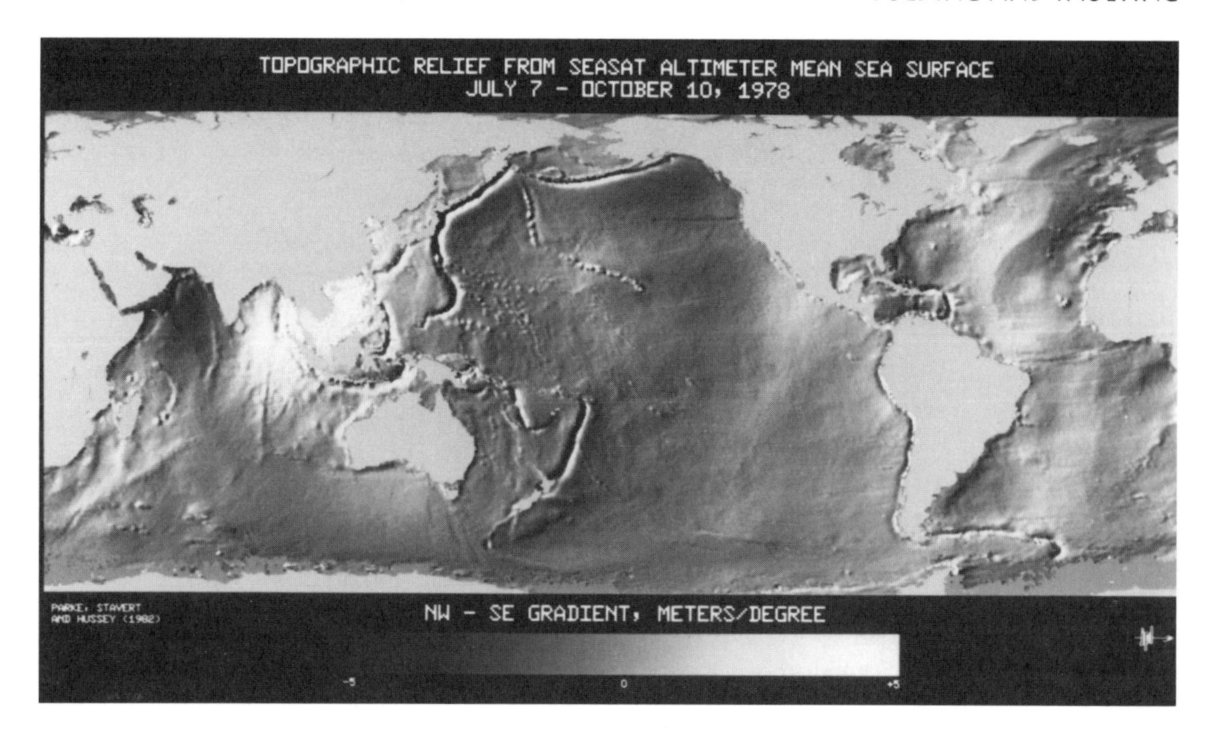

Figure 48 *Topographic relief map of the ocean surface, showing features on the ocean floor including midocean ridges and trenches.*

(Photo courtesy NASA)

As the Atlantic Basin widens, it separates the continents surrounding the Atlantic Ocean at a rate of about an inch per year, or about as fast as a fingernail grows. The spreading ocean floor in the Atlantic compresses the seafloor in the Pacific to make more room. The Pacific Basin is ringed by subduction zones (Fig. 48), which swallow oceanic plates. These are responsible for most of the geologic activity that fringes the Pacific Ocean. If placed end to end, all subduction zones would stretch all the way around the world.

The subduction of the lithosphere into the mantle plays a pivotal role in global tectonics and accounts for many geologic processes that shape the surface of the planet. The seaward boundaries of the subduction zones are marked by the deepest trenches at the edges of continents or along volcanic island arcs. Major mountain ranges and most volcanoes and earthquakes are associated with the subduction of lithospheric plates. When plates thicken and increase density, thereby losing buoyancy, they can no longer remain on the surface and sink into the mantle, forming a long line of subduction represented by a deep trench. The sinking of a plate is also the main driving force behind continental drift, and pull at subduction zones is favored over push at spreading ridges to move the continents around.

The oceanic crust is composed of basalts that originated at spreading ridges and sediments washed off the continents. When the crust along with the underlying lithosphere is subducted into the mantle, it melts and the molten rock rises toward the surface. When the magma reaches the base of the continental crust, it becomes the source material for volcanic and magmatic activity. The magma also erupts on the ocean floor, forming long chains of volcanic islands. In this manner, plate tectonics is continuously changing and rearranging the face of the Earth.

MOUNTAIN BUILDING

Mountains are areas of high relief, rising abruptly above the surrounding terrain. Folded mountain belts created by the collision of continental plates are highly deformed rocks that form the core of the range. Most mountains are built when plate motions shove the crust of one plate onto another plate. Many peaks form similar to the Wind River Mountains of Wyoming, beneath which a gently sloping fault suggests that horizontal squeezing of the continents and not vertical lifting is responsible for these ranges.

Mountains are also created when a deep root of light crustal rock literally floats the mountain like an iceberg. Additional buoyancy might be provided when the underlying lithosphere drips away from the crust and is replaced by hot rock from the mantle, further lifting the mountain range. Globs of relatively cold rock dropping hundreds of miles into the mantle appear to precede this type of mountain building. A good example is the 2.5-mile-high southern Sierra Nevada, which has risen some 7,000 feet over the last 10 million years, yet no plates have converged near the region for more than 70 million years.

Thousands of feet of sediments are deposited along the seaward margin of a continental plate in deep-ocean trenches, and the increased weight presses downward on the oceanic crust. As continental and oceanic plates merge, the heavier oceanic plate is subducted under or overridden by the lighter continental plate, forcing it farther downward. The sedimentary layers of both plates are compressed, swelling the leading edge of the continental crust. The topmost layers are scraped off the descending oceanic crust and plastered against the swollen edge of the continental crust. This process forms a mountain belt similar to the Andes of South America. In the deepest part of the continental crust, where temperatures and pressures are extremely high, rocks are partially melted and metamorphosed. Pockets of magma also provide new source material for volcanoes and other igneous activity.

Magma extruding onto the Earth's surface builds volcanic structures such as broad plateaus and mountains. The volcanoes of the Cascade Range (Fig. 49) were created by the subduction of the Juan de Fuca plate along the Cascadia subduction zone beneath the northwestern United States. As the plate melts while diving into the mantle, it feeds molten rock to magma chambers underlying the volcanoes. Besides supplying magma for this string

Figure 49 *Mount Rainier, Cascade Range, Pierce County, Washington.*

(Photo by B. Willis, courtesy USGS)

of hungry volcanoes, the subducting plate also has the potential of generating very strong earthquakes in the region.

The continental roots underlying mountain ranges can extend downward 100 miles or more into the upper mantle. Because of collisions arising from plate tectonics, continents have stabilized part of the mobile mantle rock below. The drifting continents are thus able to carry along with them thick layers of chemically distinct mantle rock. The process that forms deep roots operates by squeezing a plate into a thicker one by continental collision. Nowhere is this process more intense than from the collision between the Indian plate and the Eurasian plate, some 45 million years ago. The latter has shrunk some 1,000 miles and in the process raised the great Himalayas, the world's tallest mountains, and the Tibetan Plateau, the largest tableland in the world.

FOLDED STRATA

Before the introduction of the plate tectonics theory, the formation of mountain ranges was still very much a mystery. Geologists generally thought that mountains formed early in the Earth's history when the molten crust solidified and shriveled like a baked apple. After making more extensive studies of mountain ranges, however, they were forced to conclude that the folding of rock layers was much too intense (Fig. 50), requiring much more rapid cooling and contraction than was possible. Moreover, if mountains were formed in this manner, they would have been scattered evenly throughout the world instead of concentrated in a few chains.

Most mountains occur in ranges, and although a few isolated peaks do exist they are rare. Mountains have complex internal structures formed by folding, faulting, volcanic activity, igneous intrusion, and metamorphism. Mountain building, which supplies the forces necessary for folding and faulting rocks at shallow depths, also provides the stresses that strongly distort rocks at greater depths.

Water, wind, and ice gradually remove most signs of what were once splendid mountain ranges. However, erosion does not erase everything, and

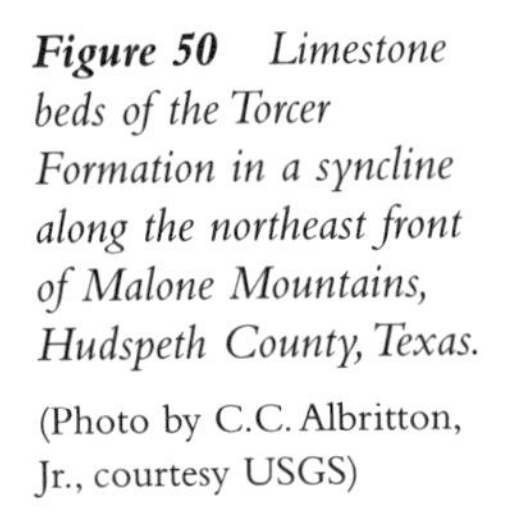

Figure 50 *Limestone beds of the Torcer Formation in a syncline along the northeast front of Malone Mountains, Hudspeth County, Texas.*

(Photo by C.C. Albritton, Jr., courtesy USGS)

Figure 51 *The Blue Ridge escarpment, Appalachian Mountains, Macon County, North Carolina.*

(Photo by A. Keith, courtesy USGS)

often the roots of ancient mountains survive beneath seemingly unimpressive landscape. Buried deep below are huge faults and folds that run through the basement rock, indicating that long ago tectonic forces squeezed the crust, causing mountains to grow. Similar folds and faults form the roots of more modern ranges such as the Rockies and the Himalayas. Vermont still preserves the roots of ancient mountains shoved upward some 400 million years ago, when the proto-North American and African continents collided.

The time required for the squeezing, heating, and alteration of rock within the new mountains remains uncertain. However, according to analysis of radioactive elements, the process of mountain folding apparently takes place in only a matter of a few million years. When continents collide, they crumple the crust and force up mountain ranges at the point of impact. The sutures joining the landmasses are preserved as eroded cores of ancient mountains called orogens from the Greek *oros* meaning "mountain." Many of today's folded mountain ranges were uplifted by Paleozoic continental collisions that raised huge masses of rocks into several mountain belts throughout the world.

The Appalachians (Fig. 51) were formed when North America and Africa slammed into each other in the late Paleozoic, during the formation of Pangaea. The southern Appalachians are underlain by more than 10 miles of sedimentary and metamorphic rocks that are essentially undeformed, whereas the surface rocks were highly deformed by the collision due to thrust faulting. Caught between the colliding continents was a proto-Atlantic Ocean called the Iapetus, which was squeezed completely dry.

About 50 million years ago, the Tethys Sea between Africa and Eurasia began to narrow as the two continental plates collided, and was closed off entirely around 20 million years ago. Like a rug thrown across a polished floor, the crust folded over into giant pleats. Thick sediments accumulating in the sea for tens of millions of years were compressed into long belts of mountain ranges on the northern and southern continental landmasses. The entire crusts of both continental plates buckled upward, forming the central portions of the ranges.

The Alps formed in a similar manner as the Himalayas when the Italian prong of the African plate plunged into the Eurasian plate when the two continents came together. The Dolomites are an impressive sight, comprising a wall of serrated peaks made of the mineral dolomite, which makes up some 10 percent of the world's sedimentary rock. It forms by the partial replacement of calcium in limestone with magnesium, making the rock substantially harder.

Additional compression and deformation might take place farther inland beyond the line of collision, creating a high plateau with surface volcanoes similar to the wide plateau of Tibet, which lies at an average elevation of more than 3 miles above sea level. The strain of raising the world's highest mountain range by the collision of the Indian plate with the Eurasia plate has resulted in deformation and earthquakes all along the plate. India is still plowing into Asia at a rate of about 2 inches a year. As resisting forces continue to build, plate convergence will eventually stop and the mountains will cease growing and begin to lose the battle with erosion.

FAULT TYPES

Faults are classified by the relationship of the rocks on one side of the fault plane with respect to those on the other side (Fig. 52). Vertical displacements along faults, with one side of the fault positioned higher than the other side, are a common form of faulting. If the crust is pulled apart, one side of the fault will slide downward past the other side along a plane that is often steeply inclined. This is known as a gravity or normal fault, a historical misnomer because this was once thought to be how faults normally occur.

Actually, most faults are produced by compressional forces, creating a reversed fault, the opposite of a normal fault, with one side of the fault pushed above the other side along a vertical or inclined plane. The great 1964 Alaskan earthquake produced as much as 50 feet of vertical displacement, forming a high scarp along the fault zone. If the reverse fault plane is nearly flat and the movement is mainly horizontal for great distances, it results in a thrust fault (Fig. 53). A thrust fault occurs when a highly compressed plate shears so that

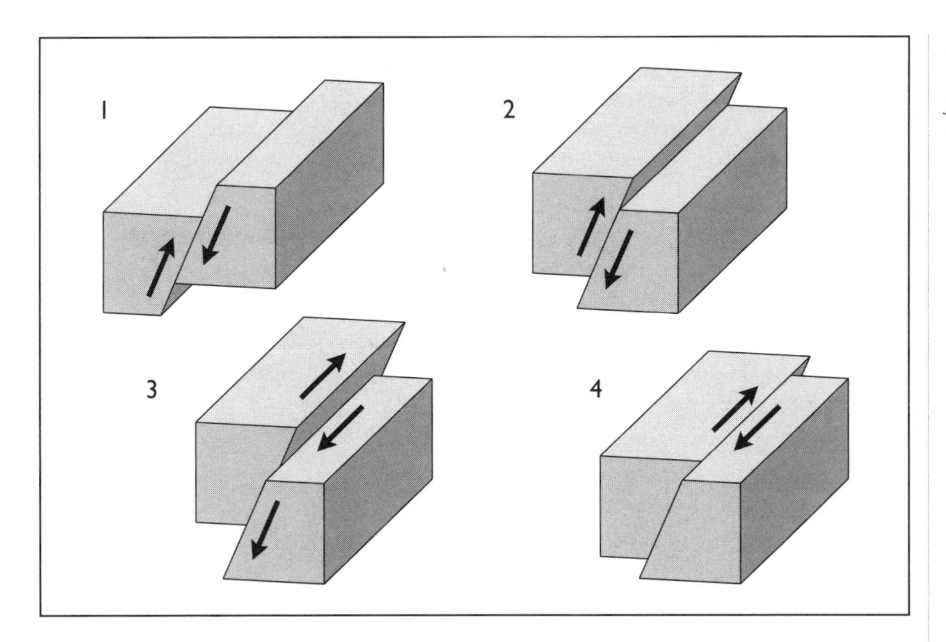

Figure 52 *Fault types: 1. normal fault, 2. reverse fault, 3. oblique fault, 4. lateral fault*

one section is lifted over another. The Overthrust Belt from Canada to Arizona is an example of this type of faulting.

When deep faults fail to break the surface during a major earthquake, it might have been caused by a blind, low-angle thrust fault. The May 2, 1983

Figure 53 *The Lewis thrust fault at the south end of Glacier National Park.*

(Photo by M. R. Mudge, courtesy USGS)

Figure 54 *A view to the west toward the Uinta Fault from Bear Mountain, Uinta Mountains, Daggett County, Utah.*

(Photo by W. R. Hansen, courtesy USGS)

Coalinga earthquake of 6.7 magnitude nearly leveled the town. The earthquake appears to have taken place on a thrust fault because of the lack of ground rupture, which should have occurred with any temblor greater than 6.0. The earthquake that struck Whittier, California, on October 1, 1987 was a little less than 6.0 in magnitude. Although the fault did not rupture the surface, damage was severe, and the hills outside town grew almost 2 inches.

Thrust faults can cause more damage than strike-slip faults for equal measures of magnitude. Strike-slip faults cause buildings to sway back and forth, and their flexible steel frames absorb most of the force. Thrust faults, on the other hand, suddenly raise and drop buildings inches at a time, creating tremendous forces that topple even the most well-designed structures.

Some faults are neither all horizontal nor all vertical, but consist of complicated diagonal movements. If a fault is a combination of both vertical and horizontal movements, it forms a complex fault system known as an oblique or scissors fault. The great Uinta Fault on the north side of the Uinta Mountains in Utah (Fig. 54) is an example of such a fault. The 1989 Loma Prieta earthquake in California (Fig. 55) ruptured a 25-mile-long segment of the San Andreas Fault.

The faulting propagated upward along a dipping plane, resulting in a right oblique reverse fault. The earthquake raised the southwest side of the fault more than 3 feet, contributing to the continued growth of the Santa Cruz Mountains.

HORSTS AND GRABENS

If a large block of crust is bounded by reverse faults and upraised with little or no tilting, it produces a long, ridgelike structure called a horst, German for "ridge" (Fig. 56). The Black Hills near Jerome in central Arizona are made from a horst with normal faults on the east and west. The hills comprise about 1,500 feet of flat-lying sedimentary strata overlying Precambrian granites. The horst is bounded on the east by the Verde Fault, with a stratigraphic throw of about 1,500 feet, and on the west by the Coyote Fault, which dips steeply westward.

If a large block of crust is bounded by normal faults and downdropped, it produces a long, trenchlike structure called a graben, German for "ditch." Grabens are generally much longer than they are wide. For example, the Rhine graben in Germany along the Rhine River Valley is 180 miles long and about 25 miles wide. Some grabens are expressed in the surface topography as linear structural depressions, with the flanking highland areas often consisting of horsts. Sometimes grabens are buried deep below ground, and the only way to find them is to drill a series of exploratory bore holes across the terrain.

Figure 55 *Collapsed building in the Marina District caused by the October 17, 1989, Loma Prieta, earthquake, San Francisco County, California.*

(Photo by G. Plafker, courtesy USGS)

Horsts and grabens are often found in association, forming long parallel mountain ranges and deep valleys, such as the Great East African Rift, Germany's Rhine Valley, the Dead Sea Valley in Israel, the Baikal Rift in Russia, and the Rio Grande Rift in the American Southwest (Fig. 57), which runs northward through central New Mexico on into Colorado.

The rift valleys in Africa are a complex system of parallel horsts, grabens, and tilted fault blocks, with a net slip on the border faults upward of 8,000 feet. The eastern rift zone lies east of Lake Victoria and extends 3,000 miles from Mozambique to the Red Sea. The western rift zone lies west of Lake Victoria and extends northward for 1,000 miles. The rift just north of Lake Victoria filled with water to form Lake Tanganyika, the second deepest lake in the world. Russia's Lake Baikal at 6,000 feet deep has the distinction of being the world's deepest lake. It fills the Baikal Rift zone, which is similar to the East Africa Rift.

Figure 56 *A down-dropped block forms a grabben (top) and an upthrust block forms a horst (bottom).*

Figure 57 *The Manzano Mountains in the background border the eastern edge of the Rio Grande Rift, Bernalillo County, New Mexico.*

(Photo courtesy USGS Earthquake Information Bulletin)

The Basin and Range Province of North America comprises numerous fault block mountain ranges bounded by high-angle normal faults. The crust in this region is broken into hundreds of pieces tilted and upraised nearly a mile above the basin, forming about 20 nearly parallel mountain ranges upward of 50 miles or more long. The region is literally being stretched apart due to the weakening of the crust by a series of down-dropped blocks.

FAULT BELTS

The vast majority of earthquakes are concentrated in a few narrow zones that wind around the globe. The area of greatest seismic activity lies on the boundaries between lithospheric plates, especially those associated with deep trenches and volcanic island arcs, where an oceanic plate is thrust under a continental plate. The greatest amount of seismic energy is released along a path near the outer edge of the Pacific Ocean, known as the circum-Pacific belt.

The circum-Pacific belt follows the subduction zones that flank the Pacific Basin and corresponds to the "Ring of Fire," the rim of the Pacific Basin that contains most of the world's active volcanoes. In the western Pacific, the belt encompasses the volcanic island arcs that fringe the subduction zones, producing the largest earthquakes in the world (Fig. 58). On the eastern side of the circum-Pacific belt, the Andes Mountain regions of Central and South America, especially in Chile and Peru, are known for the largest and most devastating earthquakes. In the 20th century, nearly two dozen earthquakes of 7.5 magnitude or greater have taken place in Central and South America. The 1960 Chilean earthquake of 9.5 magnitude is the largest recorded anywhere in the world.

The entire western seaboard of South America is affected by an immense subduction zone just off the coast. The lithospheric plate on which the South American continent rides is forcing the Nazca plate to buckle under, causing great tensions to build up deep within the crust. While some rocks are being forced deep down, others are pushed to the surface to raise the Andean mountain chain. The resulting forces are building great stresses

Figure 58 *Rescue workers survey damages at the Christian College of the Philippines in the aftermath of an earthquake in October 1990.* (Photo by Joseph Lancaster, courtesy U.S. Navy)

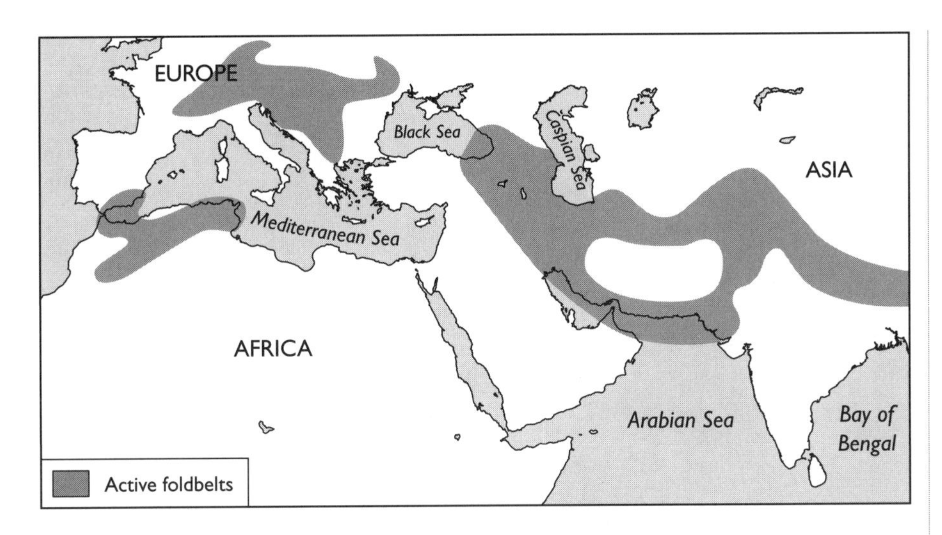

Figure 59 *Active fold belts result from crustal compression where continental tectonic plates collide at the collision of Africa with Eurasia.*

into the entire region. When the strain grows too large, earthquakes roll across the coastal regions.

Another major seismic belt runs through the folded mountains that flank the Mediterranean Sea (Fig. 59). The belt continues through Iran and past the Himalayan Mountains into China. At the eastern end of the Himalayan range lies perhaps the most earthquake-prone region in the world. An enormous seismic belt, some 2,500 miles long, stretches across Tibet and much of China. Farther west, the Hindu Kush range of north Afghanistan and nearby Tadzhikstan experience many earthquakes. From there, the Persian arc spreads through the Pamir and Caucasus Mountains and on to Turkey, which was jolted by an earthquake of 7.4 magnitude on August 17, 1999 that killed more than 17,000 people in the industrial heartland. The eastern end of the Mediterranean is a jumbled region of colliding plates, generating highly shaky ground.

EARTHQUAKE FAULTS

The mechanism behind earthquakes was poorly understood until the great 1906 San Francisco earthquake. For hundreds of miles along the San Andreas, fences and roads crossing the fault were displaced by as much as 21 feet. The San Andreas Fault is a 650-mile-long, 20-mile-deep fracture zone that runs northward from the Mexican border through southern California and represents a zone of separation between the North American and Pacific plates (Fig. 60).

During the San Francisco earthquake, the Pacific plate suddenly slid northward past the North American plate. During the 50 years prior to the earthquake, land surveys indicated displacements as much as 10 feet along the fault. Tectonic forces were slowly deforming the crustal rocks on both sides of

Figure 60 *The San Andreas Fault along Elkhorn scarp in the Carrizo Plains, San Luis Obispo County, California.*

(Photo by R. E. Wallace, courtesy USGS)

the fault, causing huge displacements. Meanwhile, the rocks were bending and storing up elastic energy. Eventually, the forces holding the rocks together were overcome, and slippage occurred at the weakest point.

The San Andreas is perhaps the best studied fault system in the world. It covers much of California, separating the southwestern part of the state from the rest of the North American continent. It is a strike-slip fault, with the segment of California west of the San Andreas along with the lithospheric plate on which it rides slipping past the continental plate in a northwesterly direction at a rate of about 2 inches per year.

The relative motion of the two plates is called right-lateral or dextral movement because an observer on either side of the fault would notice the other block moving to the right. If the two plates slid past each other smoothly, Californians would not worry so much about earthquakes. Unfortunately, the plates tend to snag, especially in the southern end of the fault and in an area known as the Big Bend in the northern part of the fault. When they attempt to tear themselves free, earthquakes rumble across the landscape. Both the 1906 San Francisco earthquake and the 1989 Loma Prieta earthquake took place on a segment of the San Andreas Fault that runs through the Santa Cruz Mountains. Because of the unsettled faults, several milder aftershocks are expected within a year after such major earthquakes.

If California were reconstructed as it was 30 million years ago, when the northern extension of the East Pacific Rise first came to intercept the North American continent, the segment west of the San Andreas Fault would have been south of the present Mexican border. If this motion continues for another 30 million years, southwest California could end south of the present Canadian border. No catastrophic earthquake would ever send southern California crashing into the sea, however. Instead, it will continue its slow journey northward, and in 50 million years the plate on which it rides will disappear down the Aleutian Trench, while the crust is plastered against Alaska.

Subsidiary faults along the San Andreas (Fig. 61) include many parallel faults such as the Hayward Fault that runs through suburban San Francisco, the Newport-Inglewood Fault, and numerous transverse faults. The Garlock Fault is a major east-trending fault. Movement along this fault is left-lateral, or sinistral, and, combined with the right-lateral movement of the San Andreas Fault, is causing the Mojave Desert to the south to move eastward with respect to the rest of California. The faults of the Mojave and adjacent Death Valley absorb about 10 percent of the total slippage between the Pacific and North American plates. The complex crustal movements associated with these faults are responsible for most of the tectonic and geologic features of California such as the Sierra Nevada and the Coast Ranges. Moreover, most of the earthquakes that plague California are produced by these faults.

Not all segments of the San Andreas Fault have ruptured in historic times. Some faults are exposed on the surface, while others are buried deep below ground, where stresses along the San Andreas increase with depth. A thrust fault that lies some 6 miles beneath the surface produced a damaging earthquake at Colinga, California, in 1983. Thrust faults associated with the San Andreas Fault system might express themselves on the surface as a series of active folds. These thrust faults occur where a strike-slip fault ends, while the relative motion between blocks of crust continues, causing one block to push past the end of the fault and slide upward along a sloping plane. Meanwhile, the block of crust being pulled away might slip downward along a normal fault.

Faults exposed on the surface generally produce deep vertical fissures, created by the movement of crustal plates in different directions. However, not all faults along which earthquakes occur are exposed, and most small earthquakes in California do not rupture the ground. Many earthquakes not associated with surface faults occur under folds, which are the product of

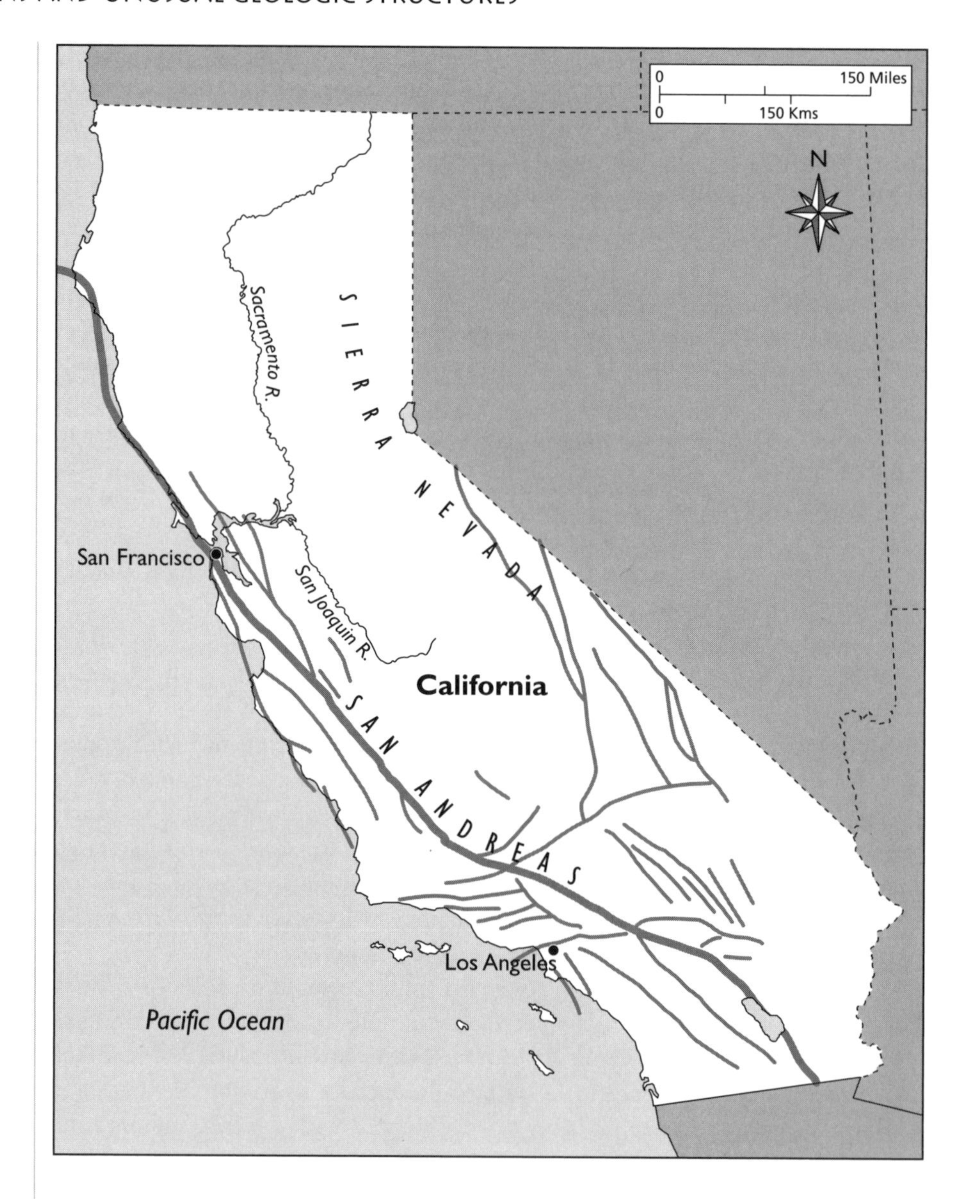

Figure 61 *The San Andreas and associated faults.*

successive earthquakes. Furthermore, folds can sometimes grow considerably during a large earthquake. For instance, an anticline (upraised strata) associated with the fault responsible for the great 1980 El Asnam earthquake in Algeria was uplifted more than 15 feet after the fault ruptured.

A fault system that resembles the San Andreas is Scotland's Great Glen Fault. It dissects the country from coast to coast and causes the highlands to the north to slide past the lowlands to the south in a left-lateral, or sinistral, direction by as much as 60 miles since the late Paleozoic. The fault trace is marked by a belt of crushed and sheared rock up to 1 mile wide. A string of deep lakes runs along the fault, including Loch Ness, famous for its mythical monster.

Another fault system that mimics the San Andreas is the 600-mile-long Red River Fault, running from Tibet to the South China Sea. When India collided with Asia some 45 million years ago, the fault allowed Indochina to slide southeastward relative to south China in a process known as continental escape. As India continued to plow into Asia, it pushed Indochina at least 300 miles to the east, forcing it to jut out to sea, rearranging the entire face of southeast Asia. The sideways escape of Indochina might have played a role in opening a new ocean basin and creating the South China Sea.

Around 20 million years ago, the fault locked and halted the continental escape. This increased the stress on Asia, which thickened the crust and raised the Himalaya Mountains and the high Tibetan Plateau. Another extremely large strike-slip fault called the Altyn Tagh runs more than 1,200 miles along Tibet's northeast border. The fault has a high rate of slip, measuring more than 1 inch per year. It is also allowing Tibet to escape to the east as India continues to plunge headlong into Asia.

Besides the San Andreas, the United States is crisscrossed by several other faults, mostly associated with mountain ranges. Most states lie in regions classified as having moderate to major seismic risk. The Basin and Range Province of southern Oregon, Nevada, western Utah, southeastern California, and southern Arizona and New Mexico, consists of many fault block mountain ranges bounded by high-angle normal faults. The crust in this region is broken into hundreds of pieces tilted and raised nearly a mile above the basin, forming nearly parallel mountain ranges upward of 50 miles or more long. The region is literally being stretched apart due to the weakening of the crust by a series of down-dropped blocks. About 15 million years ago, the locations presently occupied by Reno, Nevada and Salt Lake City, Utah were 200 to 300 miles closer together than they are today due to this stretching.

The Canadian Rockies were raised by the same mechanism of upthrusting connected with plate collision as the Andes of Central and South America. Slices of sedimentary rock were detached from the underlying basement rock and thrust eastward on top of each other. The Grand Tetons of western Wyoming (Fig. 62), one of the most spectacular mountain ranges, were upfaulted along the eastern flank and downfaulted to the west. The Wasatch Range of north-central Utah and southern Idaho is an example of a north-trending series of normal faults, one below the other, that extend for 80 miles with a net slip along the west side as much as 18,000 feet.

The upper Mississippi and Ohio River Valleys suffer frequent earthquakes, and the northeast-trending New Madrid Fault and its associated faults are responsible for three major earthquakes and many tremors. The Appalachian Mountains were formed by folding, faulting, and upwarping of

Figure 62 *Mount Moran in the Grand Teton Range, Teton County, Wyoming.*

(Photo by George A. Grant, courtesy National Park Service)

sediments and have been the seat of many earthquakes past and present. Along the eastern seaboard, major earthquakes have hit Boston, New York, Charleston, and other areas since colonial days.

EARTHQUAKES

Earthquakes are by far the most destructive, short-term natural forces on Earth. Damage arising from earthquakes is widespread, covering thousands of square miles. Not only are entire cities destroyed, but earthquakes completely change the structure of the landscape in the affected region. They can produce tall, steep-banked scarps (Fig. 63) and cause massive landslides that carry away huge blocks of earth.

Hundreds of thousands of earthquakes occur every year (Table 7), but fortunately only a few are destructive. During the 20th century, the world averaged about 18 major earthquakes of magnitudes 7.0 or larger per year. But during the last quarter of the century only about a dozen such shocks occurred on average per year. As for great earthquakes with magnitudes above 8.0, the century's average was 10 per decade. However, the number of large earthquakes is apparently on the rise.

Vertical and horizontal offsets on the surface indicate that the crust is constantly readjusting itself. These movements are frequently associated with large fractures. The greatest earthquakes are produced by sudden slippage along major faults, sometimes with offsets of several feet occurring in mere seconds. Most faults are associated with plate boundaries, and most earthquakes are generated in zones where plates are shearing past or abutting against each other.

Where plates interact, rocks at their edges are strained and deformed. This interaction can take place near the surface, where major earthquakes

Figure 63 *The Red Canyon fault scarp developed during the August 1959 Montana earthquake, Gallatin County, Montana.* (Photo by J. R. Stacy, courtesy USGS)

TABLE 7 SUMMARY OF EARTHQUAKE PARAMETERS

Magnitude	Surface Wave Height (Feet)	Length of Fault Affected (Miles)	Diameter Area Quake Is Felt (Miles)	Number of Quakes per Year
9	Largest earthquakes ever recorded—between 8 and 9			
8	300	500	750	1.5
7	30	25	500	15
6	3	5	280	150
5	0.3	1.9	190	1,500
4	0.03	0.8	100	15,000
3	0.003	0.3	20	150,000

occur, or several hundred miles below, where one plate is subducted under another. Some faulting takes place so deep it leaves no surface expression. Earthquakes are also associated with volcanic eruptions, but they are relatively mild compared with those created by faulting. Surprisingly, Antarctica and Greenland lack significant earthquakes, probably because their massive ice sheets stabilize the faults and thus inhibit fault slip.

The total amount of slippage accumulated from earthquakes over time enables scientists to estimate the velocity at which the tectonic plates bounding the fault are moving past each other. By comparing this velocity with that computed by independent geologic, magnetic, and geodetic evidence, scientists can determine how much of the plate's relative motion causes earthquakes and how much produces aseismic slip, which is ground movement unaccompanied by earthquake activity.

In areas such as Chile, noted for some of the world's largest earthquakes, all motion between plates appears to be caused by earthquake slippage alone. The massive 1960 Chilean earthquake, the largest in the 20th century, took place along a 600-mile-long rupture through the South Chile subduction zone. Usually, the longer the section of fault that breaks the larger the earthquake.

Earthquakes also occur in so-called stable zones, although not nearly as frequently as they do at plate margins. The stable zones are generally associated with continental shields, composed of ancient granitic rocks lying in the interiors of the continents and account for nearly two-thirds of all continental crust. When earthquakes occur in these regions, they might result from the weakening of the crust by compressive forces originating at plate edges.

The underlying crust also might have been weakened by previous tectonic activity, including old faults and ancient mountain belts. Failed rift systems, where spreading centers did not fully develop, are responsible for faults such as the New Madrid in the central United States, which triggered three extremely large earthquakes in the winter of 1811–12. Furthermore, the strong rock of plate interiors transmits seismic waves much more efficiently than the broken-up crust near plate boundaries. Therefore, earthquakes in these regions are felt over a much wider area. As an example, the New Madrid earthquake rang church bells as far away as Boston.

Earthquakes not associated with surface faults occur under folds, which do not rupture the Earth's surface. These earthquakes occur in many of the world's major fold belts that raise mountain chains such as those bordering the Mediterranean Sea. During the 20th century, large fold earthquakes have occurred in Japan, Argentina, New Zealand, Iran, and Pakistan. Most of these earthquakes appear to have taken place under young anticlines, which are upturned strata less than several million years old, because folds are actually the geologic product of successive earthquakes and not produced by slow creep as was once thought.

After discussing folding and faulting of the Earth's surface, the next chapter will examine the different types of igneous activity that shape the planet.

5

IGNEOUS ACTIVITY

VOLCANIC AND GRANITIC ROCKS

This chapter examines volcanoes and other types of igneous activity. The very first rocks to form on Earth were igneous, derived directly from molten magma originating from deep within the planet. New magma is generated when crustal rocks melt as they sink into the mantle at subduction zones. Magma also upwells from the asthenosphere at spreading ridges and from deep inside the mantle at hot spots. The magma slowly rises toward the surface to become the source of all volcanic and granitic activities.

This igneous activity continues to build up the continents by the addition of new rock material. Therefore, the crust of the Earth is continuously being rejuvenated, and the total mass of buoyant rocks is always preserved. The addition of new basalt on the ocean floor is responsible for the growth of oceanic crust. The movement of lithospheric plates, upon which the continental and oceanic crust ride, is responsible for the geologic forces that shape the planet.

MOLTEN MAGMA

The shifting of lithospheric plates on the surface of the planet continuously generates new crust. Deep-ocean trenches created by descending plates accumulate large amounts of sediment, primarily from the adjacent continents. The continental shelf and slope contain thick deposits of sediment washed off the continents. If the sediments are carried deep into the mantle, they melt in pockets called diapirs. The diapirs rise toward the surface to form magma bodies, which become the sources of new igneous activity (Fig. 64).

Subduction zones are noted for their active volcanism (Table 8), which builds chains of continental volcanoes and island arcs. These volcanoes produce a fine-grained, gray rock known as andesite, named for the Andes Mountains of South America. Andesite differs in composition and texture from the upwelling basaltic magma of spreading ridges. It contains a higher amount of silica, which indicates a deep-seated source, possibly as much as 70 miles below the surface.

Magma also might derive from partial melting of the subducted oceanic crust, with the heat generated by the shearing action at the top of the descending plate. Convective motions in the wedge of asthenosphere caught between the descending oceanic plate and the continental plate forces material upward, where it melts as pressure lowers.

Figure 64 *The action of seafloor spreading and subduction generates new crust.*

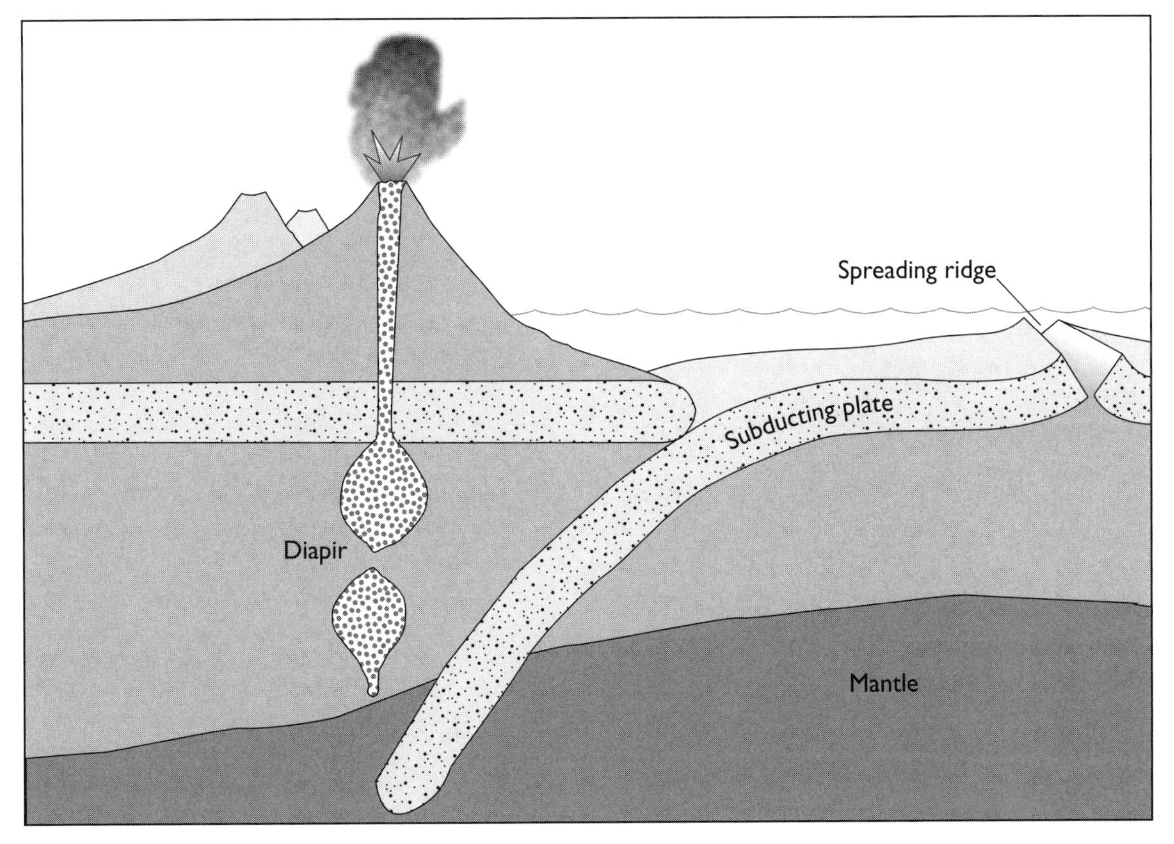

TABLE 8 COMPARISON BETWEEN TYPES OF VOLCANISM

Characteristic	Subduction	Rift Zone	Hot Spot
Location	Deep ocean trenches	Mid-ocean ridges	Interior of plates
Percent active volcanoes	80 percent	15 percent	5 percent
Topography	Mountain island arcs	Submarine ridges	Mountain geysers
Examples	Andes Mts.	Azores Is.	Hawaiian Is.
	Japan Is.	Iceland	Yellowstone
Heat source	Plate friction	Convection currents	Upwelling from core
Magma temperature	Low	High	Low
Magma viscosity	High	Low	Low
Volatile content	High	Low	Low
Silica content	High	Low	Low
Type of eruption	Explosive	Effusive	Both
Volcanic products	Pyroclasts	Lava	Both
Rock type	Rhyolite	Basalt	Basalt
	Andesite		
Type of cone	Composite	Cinder fissure	Cinder shield

The trenches, where plates dive into the mantle, are regions of low heat flow and high gravity because of the subduction of cool, dense lithosphere. Generally, the associated island arcs are regions of high heat flow and low gravity resulting from a high degree of volcanism. The back-arc basins behind island arcs are also regions of high heat flow due to the upwelling of magma from deep-seated sources.

About 80 percent of oceanic volcanism occurs along spreading ridges where magma wells up from the mantle and spews out onto the ocean floor. The spreading crustal plates grow by the steady accretion of solidifying magma to their edges. More than 1 square mile of new ocean crust, amounting to about 5 cubic miles of new basalt, is generated in this manner each year.

Seafloor spreading is often described as a wound that never heals as magma slowly oozes out of the mantle. However, at certain times, gigantic flows erupt on the ocean floor with enough new basalt to pave the entire U.S. interstate highway system 10 times over. The magma also flows from isolated volcanic structures called seamounts strung out in chains across the interior of plates. Beneath the Pacific Ocean, more than 10,000 seamounts rise from the seafloor, but only a few, such as the Hawaiian Islands and other islands crisscrossing the Pacific, manage to break the ocean surface.

The mantle material that extrudes onto the surface is black basalt, which is rich in silicates of iron and magnesium. Most of the 600 active volcanoes in the world are entirely or predominately basaltic. The magma that forms basalt originated in a zone of partial melting in the upper mantle more than 60 miles below the surface. The semimolten rock at this depth is less dense than the surrounding mantle material and rises slowly toward the surface. As the magma ascends, the pressure decreases and more mantle material melts. Volatiles such as dissolved water and gases also make the magma flow easily.

The rising magma contributes to the formation of shallow reservoirs or feeder pipes that are the immediate source for volcanic activity. The magma chambers closest to the surface are under spreading ridges where the crust is only 6 miles or less thick. Large magma chambers exist under fast spreading ridges where the lithosphere is being created at a high rate such as those in the Pacific, and narrow magma chambers exist under slow spreading ridges, such as those in the Atlantic. As the magma chamber swells with magma and begins to expand, the crest of the spreading ridge is pushed upward by the buoyant forces generated by the molten rock.

The magma rises in narrow plumes that mushroom out along the spreading ridge, welling up as a passive response to plate divergence, somewhat like taking the lid off a pressure cooker. Only the center of the plume is hot enough to rise all the way to the surface, however. If the entire plume were to erupt, it could build a massive volcano several miles high. Not all magma is extruded onto the ocean floor. Some solidifies within the conduits above the magma chamber and forms huge vertical sheets known as dikes resembling a deck of cards standing on end.

Magmas of varied composition indicate the source materials as well as the depth within the mantle from which they originated. Degrees of partial melting of mantle rocks, partial crystallization that enriches the melt with silica, and assimilation of a variety of crustal rocks in the mantle affect the ultimate composition of the magma. When the erupting magma rises toward the surface, it incorporates a variety of rock types along the way and changes composition, which is the major controlling factor in determining the type of eruption.

When the magma reaches the surface, it erupts a variety of gases, liquids, and solids. Volcanic gases mostly consist of steam, carbon dioxide, sulfur dioxide, and hydrochloric acid. The gases are dissolved in the magma and released as it rises toward the surface and pressures decrease. The composition of the magma determines its viscosity and type of eruption, mild or explosive. If the magma is highly fluid and contains little dissolved gas when reaching the surface, it flows from a volcanic vent or fissure as basaltic lava, and the eruption is usually quite mild, as with Hawaiian volcanoes (Fig. 65).

If magma rising to the surface contains a large quantity of dissolved gases, it suddenly separates into liquid and bubbles. With decreasing pressure,

Figure 65 *A high lava fountain from an eruption of Kilauea Volcano, Hawaii.*

(Photo by D. H. Richter, courtesy USGS)

the bubbles expand explosively and burst the surrounding fluid, which fractures the magma into fragments. The fragments are driven upward by the force of the expansion and hurled far above the volcano. The fragments cool and solidify during their flight and can range in size from large blocks weighing several tons to fine dust-size particles. The finer material is caught by the wind blowing across the eruption cloud and carried for long distances, sometimes completely around the world.

VOLCANIC ERUPTIONS

Volcanoes, whose eruptions are beneficial as well as hazardous, are the most spectacular of all earth processes. They are also the most important landformers, producing geologic structures ranging from a variety of cones to huge lava flows. Volcanoes can be very destructive, and often a single eruption can wipe out entire towns and take the lives of thousands of people. Volcanoes also significantly impact the climate by injecting massive amounts of volcanic ash and gases into the atmosphere.

Most volcanoes are associated with crustal movements and occur on plate margins. When one crustal plate dives beneath another, the lighter rock component melts and rises toward the surface to provide magma for volcanoes and other igneous activity. A different type of volcano created by hot spot volcanism lies in the interiors of plates and arises from magma originating deep inside the mantle, possibly from just above the core.

Volcanic islands such as the Hawaiian chain are created as the Pacific plate travels over a hot spot, as though riding on a conveyor belt. Therefore, hot spots could be a reliable means for determining the direction of plate motion. Apparently all the islands in the Hawaiian chain were produced by a single source of magma, over which the Pacific plate has passed, continuing in a northwesterly direction. Similar chains of volcanic islands exist in the Pacific that trend in the same direction as the Hawaiian Islands, including the Line and Marshall-Gilbert Islands, and the Austral and Tuamotu seamounts.

The Bermuda Rise in the western Atlantic is oriented in a roughly northeast direction parallel to the continental margin off the eastern United States. It is nearly 1,000 miles long and rises some 3,000 feet above the surrounding seafloor, where the last of the volcanoes ceased erupting about 25 million years ago. A weak hot spot unable to burn a hole through the North American plate apparently was forced to take advantage of previous structures on the ocean floor. This explains why the volcanoes trend nearly at right angles to the motion of the plate.

The Bowie seamount is the youngest in a line of submerged volcanoes running toward the northwest off the west coast of Canada. It is fed by a mantle plume nearly 100 miles in diameter and more than 400 miles below the ocean floor. However, rather than lying directly beneath the seamount, as plumes are supposed to do, this one lies about 100 miles east of the volcano. The plume could have taken a tilted path upward, or the seamount somehow moved with respect to the hot spot's position.

On the continents, hot spots leave a distinguishable trail of volcanoes. One such hot spot underlies Yellowstone National Park (Fig. 66) and can be traced across the Snake River Plain in southern Idaho. During the last 15 million years, the North American plate has traveled in a southwest direction over

Figure 66 *The Grand Canyon of the Yellowstone River, Yellowstone National Park, Wyoming.*
(Photo by F. S. Parker, courtesy USGS)

the hot spot, placing it under its temporary home at Yellowstone. In the last 2 million years, three major episodes of volcanic activity occurred in the region. The volcanic eruptions created the huge Yellowstone Caldera, about 45 miles long and about 25 miles wide. They are counted among the greatest catastrophes of nature, and another major eruption is well overdue.

Nearly all hot spot volcanism occurs in regions of broad crustal uplift or swelling where magma lies near the surface. More than half of all hot spots exist under continents. When a continental plate hovers over a number of hot spots, molten magma welling up from deep below creates domelike structures in the crust. The growing domes develop deep fissures through which magma rises to the surface. They average about 125 miles wide and account for about 10 percent of the Earth's total land area. Africa has the largest concentration of hot spots, which are responsible for the unusual topography of the African continent, characterized by basins, swells, and uplifted highlands.

Subduction zone volcanoes such as those in the western Pacific and Indonesia are among the most explosive in the world, creating new islands and destroying old ones. Their explosive nature is due to large amounts of volatiles in the magma, consisting of water and gases. When the pressure is lifted as the magma reaches the surface, these volatiles are released explosively, fracturing the magma, which shoots out of the volcano like shotgun pellets.

Indonesia is well known for its highly explosive volcanoes, including Tambora and Krakatoa, which produced the greatest eruptions in modern history (Table 9). The 1982 eruption of Galunggung and the 1983 eruption of Una Una created thick ash clouds that grounded aircraft. Alaskan volcanoes such as Mounts Katmai and Augustine, noted for their gigantic ash eruptions, resulted from the subduction of the Pacific plate at the Aleutian Trench. The June 1991 eruption of Mount Pinatubo, Philippines, possibly the largest of the 20th century, killed 700 people and left thousands homeless.

TABLE 9 THE TEN MOST WANTED LIST OF VOLCANOES

Date	Volcano	Area	Death Toll
AD 79	Vesuvius	Pompeii, Italy	16,000
1669	Etna	Sicily, Italy	20,000
1815	Tambora	Sumbawa, Indonesia	12,000
1822	Galunggung	Java, Indonesia	4,000
1883	Krakatoa	Java, Indonesia	36,000
1902	La Soufriere	St. Vincent, Martinique	15,000
1902	Pelee	Pierre, Martinique	28,000
1902	Santa Maria	Guatemala	6,000
1919	Keluit	Java, Indonesia	5,500
1985	Nevado del Ruiz	Armero, Colombia	20,000

The Cascade Range of the Pacific Northwest consists of a chain of powerful volcanoes from northern California to Canada. They are associated with the Cascadia subduction zone, which is being overridden by the North American continent. The May 18, 1980 eruption of Mount St. Helens, whose blast devastated 200 square miles of national forest, is a good example of the explosive nature of these volcanoes. One of the dirtiest volcanic eruptions of the 20th century was that of El Chichon in southeastern Mexico, which began on March 28, 1982. It sent volcanic dust clouds completely around the world, which had a major influence on the climate in the Northern Hemisphere. Mount Pinatubo had a similar effect 10 years later.

Volcanoes come in a variety of shapes and sizes. Cinder cones such as the Parícutin Volcano, which erupted in a farmer's field on February 20, 1943 (Fig. 67), form by explosive eruptions and are comparably short with steep slopes, usually less than 1,000 feet high. They are built by accumulating layer upon layer of pumice, ash, and other volcanic debris. Deep within the Earth, viscous magma contains dissolved water, carbon dioxide, and other gases. When the magma reaches the surface, the reduced pressure forces the gasses out explosively, causing the volcano to spew its contents high into the air. The debris then falls back onto the volcano, building it upward and outward.

If a volcano erupts only basaltic lava from a central vent, it forms a shield volcano. Hawaii's Mauna Loa is the largest shield volcano in the world, creat-

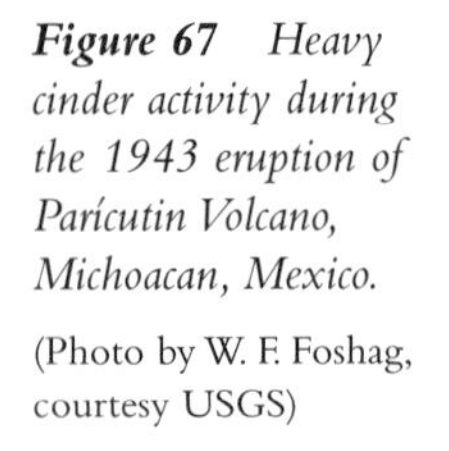

Figure 67 *Heavy cinder activity during the 1943 eruption of Parícutin Volcano, Michoacan, Mexico.*

(Photo by W. F. Foshag, courtesy USGS)

ing a great sloping dome that rises 13,675 feet above sea level (Fig. 68). Highly fluid molten rock is violently squirted out in fiery fountains of lava from pools within the crater or oozes out from a central vent. As the lava builds up in the center, it flows to the outer edge of the volcano in all directions, forming a dome-shape structure when it cools and hardens. The slope on the volcano's flanks rises only a few degrees and no more than 10 degrees near the summit. The lava spreads out to cover large areas, as much as 1,000 square miles.

Figure 68 *The broad shield volcano Mauna Loa built most of the island of Hawaii.*

(Photo courtesy USGS)

Several of these dome-shaped features in northern California and Oregon such as the Mono-Inyo Craters are 3 or 4 miles wide and 1,500 to 2,000 feet high. Lava domes grow by expansion from within because the lava is too viscous or heavy to flow very far, causing it to pile up around the vent. Lava domes commonly occur in a piggyback fashion within the craters of large composite volcanoes. Good examples are California's Mono domes and Lassen Peak.

If a volcano erupts both cinder and lava, it builds a composite volcano, also called a stratovolcano. The hardened plug in the throat of the volcano is blasted into small fragments by the buildup of pressure from trapped gases below. Along with molten rock, these fragments are sent aloft and fall back on the volcano's flanks as cinder and ash. The cinder layers are reinforced by layers of lava from milder eruptions, forming cones with a steep summit and steeply sloping flanks. These volcanoes are the tallest in the world and often end in a catastrophic collapse, forming a wide caldera. Most calderas form when a volcano loses its support and collapses into a partially emptied magma chamber. Calderas also form when a volcano blows off its upper peak, leaving behind a broad crater.

Figure 69 *The dome and crater left over from the explosive eruption of Mount St. Helens.*

(Photo by Jim Hughes, courtesy USDA Forest Service)

At the summit of most volcanoes is a steep-walled depression, or crater. The crater is connected to the magma chamber by a conduit or vent. When fluid magma moves up this pipe, it is stored in the crater until it fills and overflows. During periods of inactivity, back flow can completely drain the crater. Highly viscous lava often forms a plug in the crater, which can slowly rise to form a huge spire or dome (Fig. 69). Often, the lava is blown outward, greatly enlarging the crater.

RIFT VOLCANOES

Iceland straddles the Mid-Atlantic Ridge, where the two plates that comprise the Atlantic Basin and adjacent continents are being pulled apart. The island is a broad volcanic plateau of the Mid-Atlantic Ridge that rises above sea level and is underlain by a large mantle plume, or hot spot. The frozen island is split down the middle by a huge volcanic rift, one of the largest on land. A steep-sided, V-shaped valley runs across the island from north to south. It is flanked by many volcanoes, making Iceland one of the most volcanically active places on Earth (Fig. 70). This activity accounts for a large amount of geothermal activity, providing heat and electricity for the residents. In a geologically brief moment, Iceland will move away from its source of magma, its volcanic activity will cease, and the island will become just another ice-covered rock.

Figure 70 *Seawater is being sprayed directly on a lava flow in the outer harbor of Vestmannaeyjar, Iceland, from the May 4, 1973, eruption on Heimaey to arrest it from infilling the harbor entrance.*

(Photo courtesy USGS)

The East African Rift Valley extends from the shores of Mozambique to the Red Sea, where it splits to form the Afar Triangle in Ethiopia. Afar is perhaps one of the best examples of a triple junction created by the doming of the crust over a hot spot. The Red Sea and the Gulf of Aden represent two arms of a three-armed rift, with the third arm heading into Ethiopia. For the past 25 to 30 million years, the Afar Triangle has been stewing with volcanism and has alternated between sea and dry land. The tiny African nation of Djibouti offers the unusual phenomenon of oceanic crust being extruded as dry land. The only other site in the world where seafloor spreading can be observed on land is Iceland.

The entire African Rift is a complex system of tensional faults, indicating the continent is in the initial stages of rupture. Much of the area has been uplifted thousands of feet by an expanding mass of molten magma lying just beneath the crust. This heat source is responsible for the hot springs and volcanoes along the great rift valley. Many of the largest and oldest volcanoes in the world stand nearby, including Mounts Kenya and Kilimanjaro, the tallest mountain in Africa.

CALDERAS

The largest volcanic eruption in the continental United States in several centuries occurred on May 18, 1980, when Mount St. Helens exploded with the equivalent force of a 400-megaton nuclear bomb. The eruption blew off the top third of the mountain and created a caldera more than 1 mile wide (Fig. 71). It also produced one of the largest avalanches in recorded history and generated massive mudflows and floods that raced toward the Pacific Ocean. The blast devastated 200 square miles of forestland, and enough timber to build a fair-size city lay toppled like matchsticks.

During the last 2 million years, three major episodes of volcanic activity took place in Yellowstone National Park, Wyoming. Some 600,000 years ago, a massive eruption disgorged about 250 cubic miles of ash and pumice, equal to 1,000 times that of Mount St. Helens. The volcanic eruption created the huge Yellowstone caldera, encompassing an area of about 45 miles long and 25 miles wide. It is counted among the greatest catastrophes of nature, and another major eruption is well overdue.

Yellowstone is a typical resurgent caldera, whose floor has slowly domed upward at an average rate of about three-quarters of an inch per year since 1923. A resurgent caldera forms when the sudden ejection of large volumes of magma from a magma chamber a few miles below the surface abruptly removes the underpinning of the chamber's roof. This collapses the roof, leaving a deep, broad depression on the surface. The infusion of fresh magma into

Figure 71 *View of Mount St. Helens from the northwest following the May 18, 1980, eruption, showing massive destruction and collapse of the volcano summit, Skamania County, Washington.*

(Photo courtesy USGS)

the magma chamber causes a slow upheaval of the floor of the caldera, with a vertical uplift of several hundred feet.

If a large part of the caldera floor began to bulge rapidly at a rate of several feet a day, an eruption usually followed within a few days. As with all resurgent calderas, the one at Yellowstone formed above a mantle plume, or hot spot, that was large enough and long-lasting to melt huge volumes of rock. Resurgent calderas are recognized by widespread secondary volcanic activity such as hot springs and geysers.

Two other caldera eruptions are known to have occurred in the United States during the last million years. About 1 million years ago, a massive erup-

Figure 72 *Steam released from a geothermal well drilled into the Valles Caldera in the Jemez Plateau, Sandoval County, New Mexico.*

(Photo courtesy U.S. Department of Energy)

tion formed the Valles Caldera in northern New Mexico. A deep well was drilled into the hot depths of the dormant volcanic system to test its geothermal energy potential. An injection well was drilled on the flanks of the caldera to a depth of nearly 2 miles and encountered temperatures of 200 degrees Celsius. Water circulating through the porous rocks carried heat away from a magma chamber buried 3 miles below ground. A second recovery well brought the hot water back to the surface (Fig. 72).

Long Valley, California (Fig. 73), is a 2-mile-deep depression, resulting from a cataclysmic eruption 700,000 years ago. The caldera is just east of Yosemite National Park and measures about 20 miles long and 10 miles wide. When the volcano erupted, it fragmented the mountains that filled the area into rocky debris. About 140 cubic miles of material were strewn over a wide area, as far away as the East Coast. Magma apparently is again moving into the resurgent caldera from a depth of 7 miles below the surface. The increased volcanic activity is indicated by a rise in the center of the floor by a foot or more along with many medium-size earthquakes since 1980. If this signals an impending eruption, large portions of Nevada could be flooded with thick layers of molten rock.

As many as 10 similar eruptions have occurred in other parts of the world within the last million years. In northern Sumatra, a massive eruption 75,000 years ago created the Toba Caldera, which subsided as much as a mile or more. It is the world's largest known resurgent caldera, whose maximum dimension is nearly 60 miles. It presently contains a large lake, in which a 25-by-10-mile island was created by the upraised floor of the resurgent caldera.

Many other calderas no older than 20 to 30 million years lie in a broad belt that covers Nevada, Arizona, Utah, and New Mexico. They generally exist in zones where the crust is thinning such as rifts, where the mantle rises close

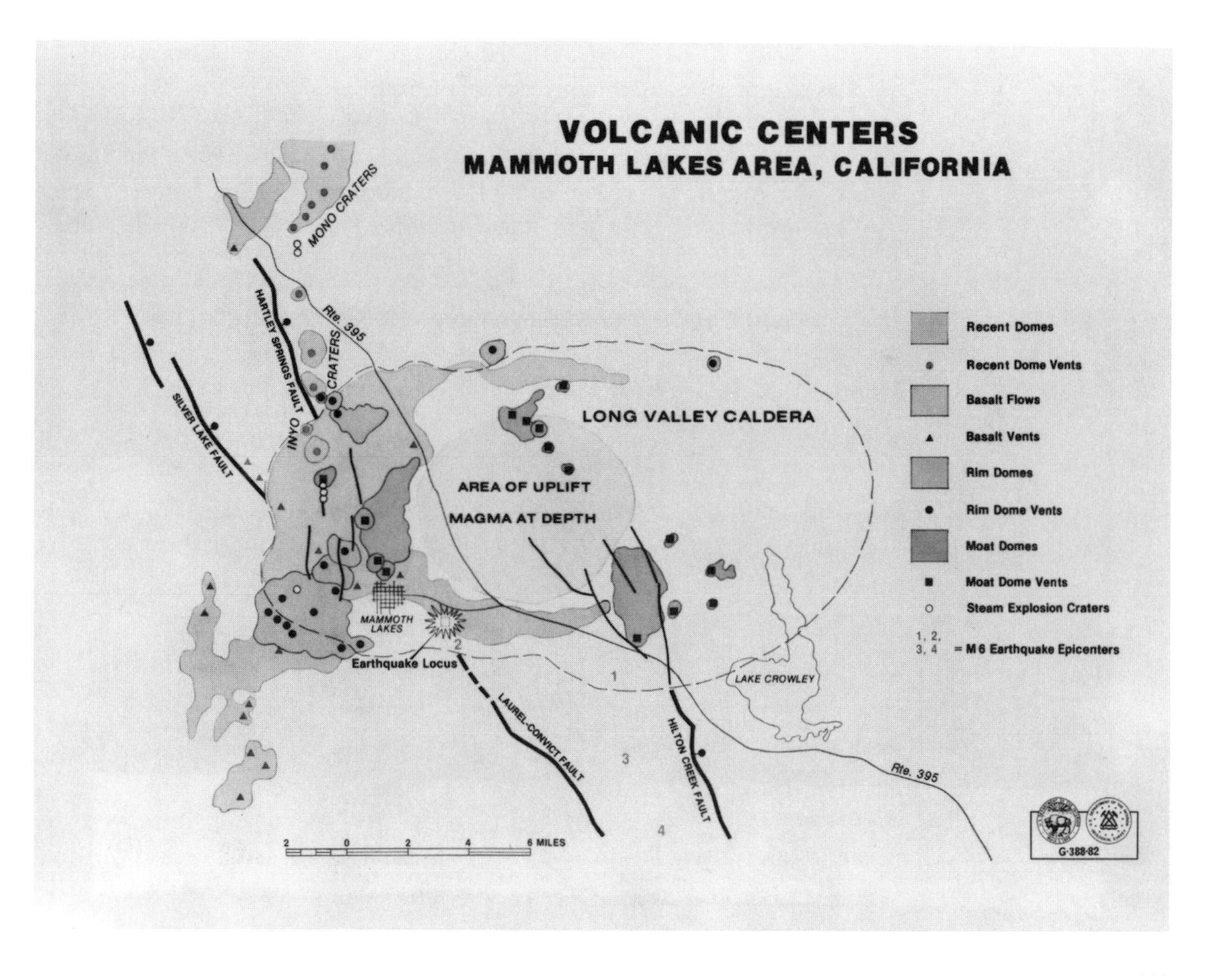

Figure 73 *A geologic map of the Long Valley Caldera in the Mammoth Lakes area, Mono County, California.* (Photo courtesy USGS)

to the surface. Calderas also form in areas where the crust has been fractured, allowing magma to move upward to the surface. The intruded magma domes pushed upward the overlying crust, creating a shallow magma chamber that contains a large volume of molten rock. The doming produces stress in the surface rock that forms the roof of the chamber and causes it to collapse along a ring fracture zone, which becomes the outer wall of the caldera after the eruption.

VOLCANIC ROCK

The products of volcanic eruptions include gases, liquids, and solids (Table 10). The main factors controlling the physical nature of volcanic products are the viscosity of the magma, its water and gas content, the rate of emission, and the environment of the vent. For example, if the vent lies underwater or beneath a glacier, the same type of magma can produce entirely different rock types due to the different cooling rates. Basalt erupted beneath glacial ice produces volcanic rocks called hyaloclastics, which are pillow lavas and pillow breccias that are unique quickly frozen forms of lava.

Many subduction zone or island arc volcanoes have higher concentrations of gas in the upper parts of their magma chambers before they erupt. This is why Indonesian volcanoes such as Tambora and Krakatoa are so explosive. The eruption begins with the emission of pyroclastics, which literally

TABLE 10 CLASSIFICATION OF VOLCANIC ROCKS

Property	Basalt	Andesite	Rhyolite
Silica content	Lowest about 50%, a basic rock	Intermediate about 60%	Highest more than 65%, an acid rock
Dark mineral content	Highest	Intermediate	Lowest
Typical minerals	Feldspar	Feldspar	Feldspar
	Pyroxene	Amphibole	Quartz
	Olivine	Pyroxine	Mica
	Oxides	Mica	Amphibole
Density	Highest	Intermediate	Lowest
Melting Point	Highest	Intermediate	Lowest
Molten rock viscosity at the surface	Lowest	Intermediate	Highest
Formation of lavas	Highest	Intermediate	Lowest
Formation of pyroclastics	Lowest	Intermediate	Highest

Figure 74 *Pyroclastic flow deposits at the base of Mount St. Helens from the October 17, 1980 eruption, Skamania County, Washington.*

(Photo courtesy USGS)

means fire fragments (Fig. 74). This is followed by thick, viscous lava flows. The texture of pyroclastics and lava is largely controlled by the number and size of gas bubble holes, called vesicles, formed in the erupting material. Pumice, the lightest of volcanic materials, contains the largest number of vesicles and can literally float on water.

Basalt, the densest volcanic rock, is formed at high temperatures and has practically no vesicles. It is the commonest rock formed from solidifying magma extruded on the surface of the Earth, moon, and many other bodies in the solar system. Sometimes, especially on the ocean floor, basalt solidifies into elongated masses called pillow lava (Fig. 75). As basalt lava cools on the surface, it shrinks, resulting in cracking or jointing. The cracks shoot vertically through the entire lava flow, breaking it into polygonal pillars or columns over a foot across.

All solid particles ejected into the air from volcanic eruptions are known as *tephra,* from the Greek word for "ash," a historical misnomer leftover from the days when volcanoes were thought to arise from the burning of subterranean substances. Tephra includes an assortment of fragments from large

Figure 75 *Pillow lava on Knight Island, Alaska.*

(Photo by F. H. Moffit, courtesy USGS)

blocks the size of automobiles to dust-size material. It forms when molten rock containing dissolved gas rises through a conduit and suddenly separates into liquid and bubbles as it nears the surface. With decreasing pressure, the bubbles grow larger. If this event occurs near the orifice, a mass of froth might spill out and flow down the sides of the volcano.

If this reaction occurs deep down in the throat of the volcano, the bubbles expand explosively and burst the surrounding liquid, fracturing the magma into fragments. These are driven upward and hurled high above the volcano. The fragments cool and solidify during their flight. Often they whistle as they gyrate wildly through the air. Blobs of still-fluid magma, called volcanic bombs, might splatter the ground nearby (Fig. 76). If they cool in flight, they form a variety of shapes, depending on how fast they are spinning. If the bombs are the size of a nut, they form *lapilli* from Latin meaning "little stones," which form strange gravel-like deposits after they land.

Tephra that flows down the slopes of a volcano supported by a layer of hot gases is called a *nuée ardente,* from the French meaning "glowing cloud." The cloud of ash and pyroclastics flows streamlike near the ground and might follow existing river valleys for tens of miles at speeds upward of 100 miles per hour. The best-known example was the 1902 eruption of Mount Pelée, Martinique, which in minutes annihilated 30,000 inhabitants.

When the tephra cools and solidifies, it forms deposits called ash-flow tuffs that can cover an area up to 1,000 square miles or more. Volcanoes also provide many samples of diverse welded tuffs, agglomerates, and ignimbrites. Large ignimbrite sheets are composed of layers of welded or recrystallized volcanic ash. For example, huge ignimbrite sheets composed of solidified deposits of volcanic ash cover the Altiplano region of the Andes in South America.

Lava is molten magma that reaches the throat of a volcano or the top of a fissure vent without exploding into fragments and flows onto the surface. The magma that produces lava is more fluid than that which produces tephra. This allows volatiles and gases to escape more easily and produces much quieter and milder eruptions such as those of Kilauea on the main island of Hawaii. Lava is mostly composed of basalt, which comprises about 50 percent silica, is dark in color, and is quite fluid.

The outpourings of lava come in two general classes, which have Hawaiian names and are typical of Hawaiian eruptions. *Pahoehoe,* or ropy lavas, are highly fluid basalt flows formed when the surface of the flow congeals to form a thin plastic skin. *Aa* or blocky lavas form when viscous, subfluid molten rock presses forward, carrying a thick, brittle crust along with them.

Figure 76 *A volcanic bomb, which fell on the east side of the Kilauea Volcano during the 1959–1960 eruption.*

(Photo courtesy National Park Service and USGS)

***Figure* 77** *An aa lava flow entering the sea from the January 21, 1960 eruption of Kilauea, Hawaii.*

(Photo by D. H. Richter, courtesy USGS)

As the lava flows, it stresses the overriding crust, breaking it into rough, jagged blocks carried along with the flow in a disorganized mass (Fig. 77).

Ancient lava flows might contain clear, dark green, or black natural glass called obsidian. Some lava flows might contain cavities filled with crystals called zeolites, meaning boiling stones, formed when water boiled away as the basalt cooled. Trachytes often possess large, well-shaped feldspar crystals aligned in the direction of the lava flow.

GRANITIC INTRUSIVES

Magma bodies that invade the crust assimilate the surrounding rocks as they melt toward the surface. This produces two major classes of igneous rocks: intrusives, which are derived from the invasion of the crust by a magma body, and extrusives, which are derived from the eruption of magma onto the surface. Because their source materials are much the same, both types of rocks share similar chemical compositions, but have different textures due to differences in cooling rates.

Magma that pours out onto the surface cools much faster than that which remains in the crust and therefore produces rocks with finer crystals. Intrusive bodies take a great deal of time to cool, perhaps a million years or more because the rocks they invade are good insulators and hold in the heat. The magma body is thus able to segregate into various components, allowing large crystals to grow. Generally, the larger the magma body the longer it takes to cool and consequently the larger the crystals.

Intrusive magma bodies called plutons exist in a variety of shapes and sizes. The largest plutons are batholiths, which are larger than 40 square miles on their exposed surfaces and are usually much longer than they are wide. Batholiths produce major mountain ranges such as the Sierra Nevada in California, which is nearly 400 miles long but only about 50 miles wide, and the Idaho Batholith, which is 250 miles long and 100 miles wide. Batholiths consist of granitic rocks with large crystals, composed mostly of quartz, feldspar, and mica. An intrusive magma body shaped much like a batholith but smaller than 40 square miles on its exposed surface is called a stock. A stock might also be a projection of a larger batholith deeper down. Like a batholith, it is composed of coarse-grained granitic rocks.

If an intrusive magma body is tabular in shape and considerably longer than it is wide, it results in a dike. A dike forms when magma fluids occupy a large crack or fissure in the crust. Because dike rocks are usually harder than the surrounding material, they generally form long ridges when exposed by erosion (Fig. 78). Sills are similar to dikes in their tabular form, but are produced parallel to planes of weakness such as sedimentary beds. A special type of sill called a laccolith bulges the overlying sediments upward, sometimes creating isolated mountain peaks in the middle of nowhere.

KIMBERLITE PIPES

Kimberlite pipes, named for the South African town of Kimberley, are the cores of ancient extinct volcanic structures that extend deep into the upper mantle, as much as 150 miles or more beneath the surface, and have been exposed by erosion. Most known kimberlite pipes were emplaced during the Cretaceous period between 135 and 65 million years ago. They brought diamonds formed billions of years ago from the upper mantle to the surface and are mined extensively for these gems throughout Africa and other parts of the world. Most economic kimberlite pipes are cylindrical or conical structures up to a mile across.

The pipes are not only important for supplying the world with diamonds, but they also sample part of the upper mantle as no other volcanic structure can. At great depth, high temperatures and pressures convert the crystal structure of carbon into a tight lattice, forming the hardest known substance on Earth. Associated with the diamonds are ultramafic (high magnesium and iron) nodules that resemble rounded cobbles. The nodules were brought up from great depths along with the diamonds and are just as rare. The large proportion of peridotite in the nodules suggests that this mineral comprises a major constituent of the mantle. In addition, xenoliths, from Greek meaning "foreign rocks," were among the mantle rocks. They were torn loose from the walls of the volcanic pipe during explosive eruptions, which in turn widened the orifice of the pipe.

The intrusive bodies vary in size and width, and most are roughly circular and pipe-shaped. More than 700 kimberlite pipes and other intrusive structures have been uncovered in South Africa, although only a few contain minable grades of diamonds. The kimberlite deposits were originally worked as open pits, but as the mines descended deeper underground, mining methods were employed. At the Kimberley mine, the world's deepest diamond mine, the diameter of the pipe at the surface is about 1,000 feet and decreases sharply with depth. Mining ceased in 1908 at a depth of 3,500 feet due to flooding, even though the diamond-bearing pipe continued to greater depths.

In North America, a concentration of kimberlite pipes lies along the border between Colorado and Wyoming, and others are known to exist in

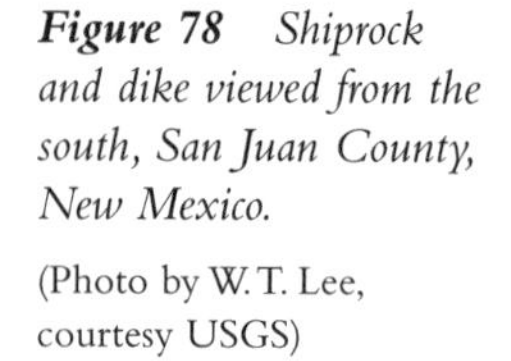

Figure 78 *Shiprock and dike viewed from the south, San Juan County, New Mexico.*

(Photo by W. T. Lee, courtesy USGS)

Montana and in the Canadian Arctic. Few North American pipes are large or economical, except a solitary pipe near Murfreesboro, Arkansas, which was briefly worked as a diamond mine beginning in 1906 and produced a total of about 40,000 stones (Fig. 79). The mine has since been turned into a tourist attraction, known as Crater of Diamonds State Park, where people pay for the privilege of sifting through the black volcanic dirt in search of that elusive stone.

Figure 79 *Hydrologic mining at the Arkansas diamond mine south of Murfreesboro, Arkansas, in 1923, Pike County, Arkansas.*

(Photo by H. D. Miser, courtesy USGS)

IGNEOUS ORE BODIES

For more than 200 years, geologists have conducted a lively debate on the origin of mineral ore deposits. The Neptunists hold that all mineral deposits were derived from water penetrating downward. The Plutonists, on the contrary, hold that mineralization was derived from the expulsion of magmatic volatiles migrating upward. These magmatic and hydrothermal processes are important in producing much of the world's wealth.

Mineral ore deposits form very slowly, taking several millions of years to create an ore rich enough to be suitable for mining. Copper, tin, lead, and zinc

Figure 80 *Steam fumaroles at Steamboat Springs, Nevada.*

(Photo by W. D. Johnston, courtesy USGS)

ores are concentrated directly by magmatic activity, especially by the intrusion of magma bodies into the Earth's crust. These concentrations are formed as hydrothermal vein deposits, which are mineral fillings precipitated from hot waters percolating along underground fractures.

Around the turn of the 20th century, geologists found that hot springs at Sulfur Bank, California, and at Steamboat Springs, Nevada (Fig. 80), were depositing the same metal-sulfide compounds that are found in ore veins. Therefore, if the hot springs were depositing ore minerals at the surface, then hot water might be filling fractures in the rock with ore as it moves toward the surface.

After excavating the ground a few hundred yards from Steamboat Springs, the American mining geologist Waldemar Lindgren discovered rocks with the texture and mineralogy of typical ore veins. He proved that many ore veins formed by circulating hot water known as hydrothermal fluids. This concept vastly improved mineral exploration because any evidence of hydrothermal alteration of rocks on the surface was enough to focus attention on the area. Unfortunately, only a few hydrothermal areas actually contain minable ore deposits. Therefore, another process must be at work.

Because most metals are found as sulfides, a source of sulfur and the chemistry that makes metal sulfides stable is needed. Geologists are still uncertain how hot water can carry enough metal to its place of deposition due to extremely low concentrations of these metals in solution. Either an ore deposit requires a huge amount of water over a very lengthy period, or some hot waters can carry more metal than what is observed on the surface.

Possibly the rocks surrounding the magma chamber are the true source of the minerals found in hydrothermal veins. In this case, the volcanic rocks only act as a heat source that pumps groundwater into a giant circulating system. Cold, heavier water moves down and into the cooling volcanic rocks carrying trace quantities of valuable elements leached from the surrounding rocks. When heated by the cooling magma body, the water becomes less dense and rises into the fractured rocks above. After cooling and losing pressure, the water precipitates its mineral load into veins and moves down again to pick up another load of minerals.

A gigantic underground still is supplied with heat and some ingredients from magma chambers. As the magma cools, silicate minerals such as quartz crystallize first, leaving behind a concentration of other elements in a residual melt. Further cooling of the magma causes the rocks to shrink and crack, allowing the residual magmatic fluids to escape toward the surface and invade the surrounding rocks, where they form veins. Certain minerals precipitate over a wide range of temperatures and pressures. This is why they are commonly found with one or two of the minerals predominating in sufficiently high concentrations to make their mining profitable.

After an examination of the basic building blocks of the Earth's crust, the next chapters will focus attention on specific types of rock formations and geologic structures.

6

CANYONS, VALLEYS, AND BASINS

DEPRESSIONS IN THE EARTH

This chapter examines some major depressions sculptured into the Earth, including canyons, rift zones, trenches, valleys, and basins. Some of the most impressive scenery the planet has to offer is carved out of solid rock by water in motion. Erosion gouges out the deepest ravines, cuts down the tallest mountains, and obliterates most other geologic structures.

Perhaps the best place to witness the power of erosion is the Grand Canyon of the Colorado River. The ocean bottom rivals the land for its rugged topography, and many ravines on the seafloor can hold several Grand Canyons. The most monumental landforms are carved out of the Earth by massive glaciers. Rivers and streams also play a major role in sculpting the landscape, providing a cornucopia of unusual geologic structures.

TERRESTRIAL CANYONS

Among the most magnificent examples of the power of erosion are canyons. They are generally found in arid or semiarid regions, where the effect of stream erosion is greater than that of weathering. Arizona's Grand Canyon (Fig. 81) is a great gash in the earth 277 miles long, 10 miles wide on average, and more than 1 mile deep. It lies at the southwest end of the Colorado Plateau, a relatively mountain-free expanse that stretches from Arizona north into Utah and east into Colorado and New Mexico.

Initially, the area surrounding the canyon was almost totally flat. Over the last 2 billion years, heat and pressure buckled the land into mountains that were later flattened by erosion. Again, mountains formed and were eroded and flooded by shallow seas. Afterward, the land was uplifted during the growth of the Rocky Mountains between 80 million and 40 million years ago.

The Grand Canyon is a fairly young feature. It is a giant cut in the Earth's crust formed by forces of uplift and erosion as the Colorado River

Figure 81 *A view of the Grand Canyon from Toroweap Point.*

(Photo courtesy National Park Service)

Figure 82 *The upper horizontal Plateau series is separated by an angular unconformity from the older, tilted Grand Canyon series.*

(Photo by L. F. Noble, courtesy USGS)

carved a path through some 500 million years of accumulated sediments and Precambrian basement rock. About 10 to 20 million years ago, the Colorado River began eroding layers of sediment along the Colorado Plateau. Its present course was established between 5 and 6 million years ago, when Baja California separated from the Mexican mainland, providing a new outlet to the sea. Much of the canyon was not carved out by piecemeal erosion grain by grain, but by catastrophic landsliding.

The canyon provides some of the best rock exposures on the North American continent. It slices through rocks hundreds of millions of years old and more than a mile thick. Well-exposed sedimentary layers tell a near complete story about the geologic history of the area. On the bottom of the canyon lies the ancient original basement rock, upon which sediments were slowly deposited layer by layer. On the canyon wall is a unique angular unconformity separating the upper horizontal Plateau series from the older, tilted Grand Canyon series (Fig. 82).

The Great Unconformity, readily exposed in the Grand Canyon, is one of North America's most prominent and famous geologic features. It stretches across much of the continent from Arizona to Wisconsin to Alberta, Canada. The Great Unconformity is a boundary where relatively young rocks lie atop much older rocks, in some places more than 2 billion years older. It marks a gap in time where a sequence of rock layers was scoured away by erosion and then new sediments were deposited in its place. Just above the unconformity lies a layer of sandstone deposited along an ancient shoreline. The sandstone varies in age across North America, with the youngest sediments lying near the center and the oldest near the edges of the continent.

Across North America, the Great Unconformity is considered a single geologic feature because the top of the ancient bedrock that underlies it was once the original surface of the Earth. Erosional processes scoured the bedrock, creating a flat surface upon which sediments were later deposited. In some places, old rock bulges up into the overlying layers of younger rock, marking the sites of hills in the ancient landscape. In other areas, fragments of the original bedrock are imbedded in layers immediately above the unconformity, indicating periods of erosion and redeposition.

The Great Unconformity is easily recognized in the Grand Canyon, where the roaring Colorado River cuts into the crust. Layers of sedimentary rock are piled high and arranged in orderly horizontal layers, with the youngest strata on top and the oldest on the bottom. At 4,000 feet below the rim of the canyon, however, the rock abruptly changes. Here, the oldest sediments, which are only 540 million years old, sit atop dark metamorphic rock some 2 billion years old, creating a gap in time of nearly 1.5 billion years.

Nearby, in the Colorado National Monument near Grand Junction, Colorado, the sandstone found elsewhere at the upper edge of the Great Unconformity is replaced by 200-million-year-old rock of the Chinle Formation (see cross section of the Grand Canyon in Chapter 3). Apparently, the sandstone in this region eroded away long before the Chinle was deposited. This probably occurred while the so-called ancestral Rocky Mountains were being raised some 300 million years ago. Therefore, the Great Unconformity represents an even greater missing period of geologic history than the gap in time at the Grand Canyon.

Among the best exposures of Precambrian rocks in the United States are the 1.8-billion-year-old metamorphic rocks on the bottom of the Grand Canyon (Fig. 83). More than a mile of sedimentary rocks overlie the bedrock of the Grand Canyon. During this time, the floor of the Grand Canyon was worn down by erosion, creating a gap in time known as a hiatus. Thick deposits of marine sediments were slowly laid down on the Grand Canyon floor. The continuous buildup of sediments caused the ancient seafloor to subside due to the increased weight.

In a fraction of the time required to deposit the sediments, a gradual upheaval brought them to their present elevation, while the Colorado River gouged out layer upon layer of rock, exposing the raw earth below. The Imperial Valley of southern California owes its rich soil to the Colorado River as it carved out the mile-deep Grand Canyon and deposited its sediments in the area to a depth of 3 miles.

Figure 83 *The Precambrian Vishnu Schist on the bottom of the Grand Canyon.*

(Photo by R. M. Turner, courtesy USGS)

SUBMARINE CANYONS

On the floor of the ocean lies a rugged landscape that rivals the land. Chasms plunge to depths that dwarf even the largest canyons on the continents. Several deep canyons slice through the continental shelf beneath the Bering Sea between Alaska and Siberia. About 75 million years ago, continental movements created the broad Bering Shelf rising 8,500 feet above the deep ocean floor. Several times the shelf was exposed as dry land, during the ice ages when sea levels dropped several hundred feet. Terrestrial canyons cut deep into the shelf. When the ocean refilled again at the end of the last ice age, massive landslides and mudflows swept down steep slopes on the shelf's edge, gouging out 1,400 cubic miles of sediment and rock.

During the last ice age, about 10 million cubic miles of the Earth's water were held in the continental ice sheets. Glaciers covered about a third of the land surface with an ice volume three times greater than its present size. The accumulated ice dropped the level of the ocean about 400 feet, exposing land bridges and linking continents. The drop in sea level advanced the shoreline hundreds of miles seaward. The coastline of the eastern seaboard of the United States extended about halfway to the edge of the continental shelf, which runs eastward more than 600 miles.

A step on the continental shelf off the eastern United States has been traced for nearly 200 miles. It appears to represent the former ice age coastline, now completely submerged. Submarine canyons slice through the continental margin and ocean floor off eastern North America. The massive continental glaciers that sprawled over much of the Northern Hemisphere held enough water to lower the sea by several hundred feet. Rivers flowing across the exposed land gouged out several canyons in the ocean floor. Submarine canyons carved into bedrock 200 feet below sea level can be traced to rivers on the continent. They were carved by rivers emptying into the sea, when the sea level was lowered dramatically during the last ice age.

Submarine canyons on continental shelves and slopes possess many identical features as river canyons, and some rival even the largest on the continents. They are characterized by high, steep walls and an irregular floor that slopes continually outward. The canyons are up to 30 miles and more in length, with an average wall height of about 3,000 feet. Some submarine canyons were carved out of the ocean floor by ordinary river erosion during a time when sea levels were much lower than they are today. Many submarine canyons have heads near the mouths of large rivers. The Great Bahamas Canyon is one of the largest submarine canyons, with a wall height of 14,000 feet, over twice as deep as the Grand Canyon.

Some submarine canyons extend to depths of more than 2 miles, making them too deep for a terrestrial river origin. They formed instead by undersea slides, which carve out deep gashes in the ocean floor. The slides move rapidly down steep continental slopes covered mainly with fine sediments swept off the continental shelves by these slides. The slides consist of sediment-laden water that is heavier than the surrounding seawater. The turbid water moves swiftly along the ocean floor and effectively erodes the soft bottom material. These muddy waters, called turbidity currents, can move down the gentlest slopes and transport huge portions of the ocean floor.

CONTINENTAL RIFTS

The continental lithosphere, the solid outer shell of the Earth, is generally between 50 and 100 miles thick. Cracking open a continent seems to be a formidable task due to its great thickness. Somehow, when continents rift apart into separate plates, thick lithosphere must somehow give way to thin lithosphere. The transition from a continental rift to an oceanic rift is preceded by block faulting. Blocks of continental crust drop down along extensional faults where the crust is being pulled apart, resulting in a deep rift valley and a thinning of the crust.

The rifting of a continent begins with hot-spot volcanism at rift valleys. The hot spots act like geologic blowtorches that burn holes through the crust and weaken it. The hot spots are connected by rift valleys, along which the continent eventually splits apart. Mantle material rises in giant plumes and under-plates the crust with basaltic magma. This further weakens it, causing huge blocks to downdrop and form a series of grabens, or fault-bounded trenches.

The rising mantle convection currents spread out in opposite directions beneath the lithosphere and pull the thinning crust apart, forming a deep rift valley. In the process of rifting, large earthquakes strike the region as huge blocks of crust drop down along diverging faults. In addition, massive volcanoes erupt due to the abundance of molten magma rising up from the asthenosphere. The crust beneath the rift is only 20 to 30 miles thick compared with 50 miles or more thick for the rest of the continent. As the crust continues to thin, magma chambers rise closer to the surface and volcanic eruptions become more prevalent. A marked increase in volcanism produces vast quantities of lava that flood onto the continental crust during the early stages of many rifts.

When the continent fragments, the rift valleys spread farther apart and eventually become a new ocean with an oceanic rift as seawater flows in and floods the region. As the rift continues to widen and deepen, it is replaced by a spreading ridge system, where hot mantle material wells up through the rifts to form new oceanic crust between the two separated segments of continental crust. More than 2 million cubic miles of molten rock are released when a supercontinent rifts apart over a hot-spot plume, which explains the large increase in volcanic activity during the early stages of continental rifting.

Excellent evidence for the rifting of continents is found at the great East African Rift Valley (Fig. 84), which extends from the shores of Mozambique to the Red Sea, where it splits to form the Afar Triangle in Ethiopia. When it finally breaks up, the continental rift will be replaced by an oceanic rift. This process is presently taking place in the Red Sea (Fig. 85), which is rifting from south to north. The Gulf of Aden is a young oceanic rift between the ruptured continental blocks of Arabia and Africa, which have been diverging for more than 10 million years. The breakup of North America and Europe beginning about 170 million years ago might have been accomplished by the same upwelling of basaltic magma that is now occurring under the East African and Red Sea Rifts.

OCEANIC RIFTS

Rifts on the seafloor form midocean spreading ridge systems. The floor of the Atlantic Ocean acts as a huge conveyor belt that transports lithosphere outward from its point of origin at the Mid-Atlantic Ridge (Fig. 86). The Mid-Atlantic Ridge bisects the Atlantic nearly exactly down the middle, weaving halfway between continents and assuming the shapes of continental margins on opposite shorelines. It is the most extraordinary mountain range on Earth.

The submerged mountains and undersea ridges form a continuous 45,000-mile-long chain several hundred miles wide and up to 10,000 feet high.

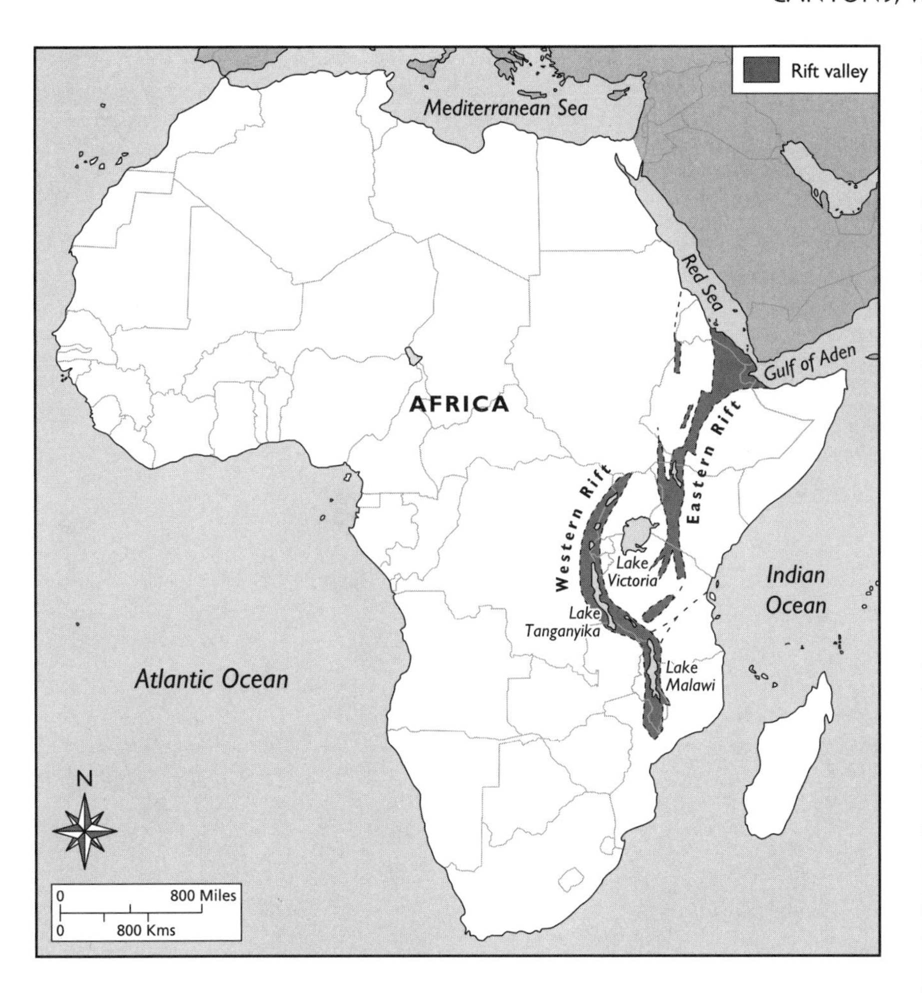

Figure 84 The East African Rift system, Red Sea, and Gulf of Aden.

Although it is deeply submerged, the spreading ridge system is easily the most dominant feature on the planet, extending over an area larger than all major terrestrial mountain ranges combined. Running down the middle of the 10,000-foot-high ridge crest is a deep trough like a giant fissure in the ocean's crust. The trough is 4 miles deep in places, or four times deeper than the Grand Canyon, and up to 15 miles wide, which qualifies it as the largest canyon on Earth.

The axis of the Mid-Atlantic Ridge is offset laterally in a roughly east-west direction by transform faults, which comprise a deep trough joining the tips of two segments of the ridge. Friction between segments produces strong shearing forces that wrench the ocean floor into steep canyons. Fracture zones offset the axis of the ridge, the largest of which is the Romanche Fracture Zone in the equatorial Atlantic. It extends for nearly 600 miles and has a vertical relief of 4 miles. The fracture zone is flanked by several similar zones, producing a sequence of troughs and transverse ridges.

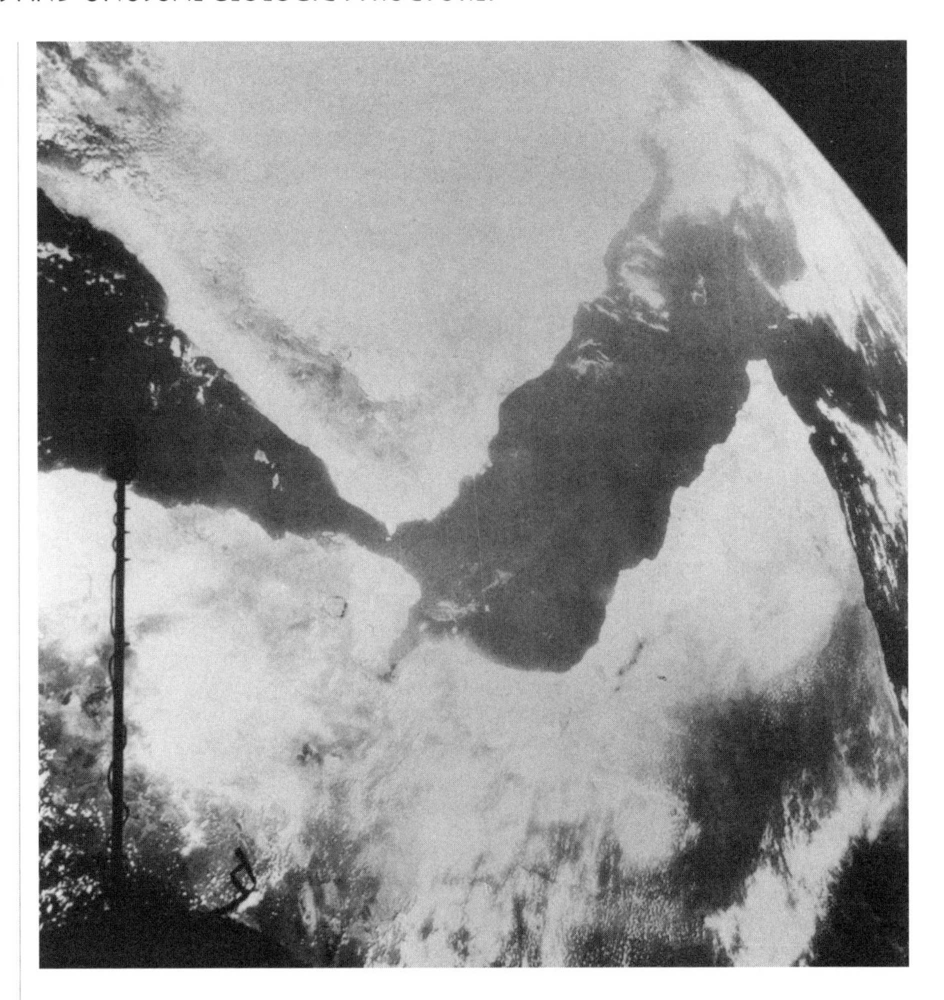

Figure 85 *The Red Sea and the Gulf of Aden are prototype seas created by seafloor spreading.*

(Photo courtesy USGS Earthquake Information Bulletin)

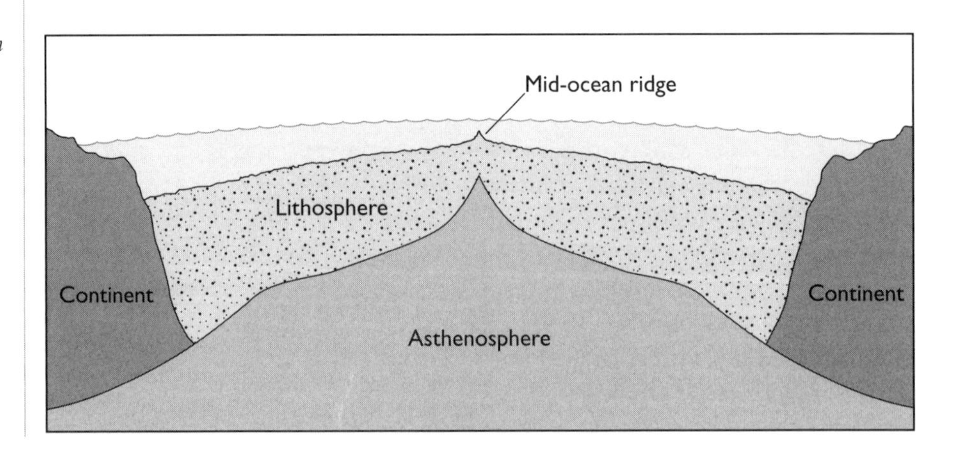

Figure 86 *Cross section beneath the spreading Mid-Atlantic Ridge.*

Figure 87 *The rim of a lava lake collapse pit on the Juan de Fuca Ridge in the East Pacific.*

(Photo courtesy USGS)

The ocean floor at the crest of the midocean ridge consists almost entirely of hard volcanic rock. The ridge system exhibits many unusual features, including massive peaks, saw-tooth ridges, earthquake-fractured cliffs, deep valleys, and unusual lava formations (Fig. 87). Along much of its length, the ridge system is carved down the middle by a sharp break or rift that is the center of an intense heat flow. In addition, the spreading ridges are the sites of frequent earthquakes and volcanic eruptions, as though the entire system were a series of giant cracks in the Earth's crust.

The East Pacific Rise is a 6,000-mile-long rift system along the eastern edge of the Pacific plate. It is the counterpart of the Mid-Atlantic Ridge and a member of the world's largest mountain chain (Fig. 88). The rift system is a network of midocean ridges, most of which lies underwater. Each rift is a narrow fracture zone, where plates of the oceanic crust are being pulled apart by the action of plate tectonics.

On the crest of the East Pacific Rise, at the base of jagged basalt cliffs more than 2 miles deep are lava flows and fields strewn with pillow lava. Active hydrothermal fields lie in areas where seawater seeping downward near magma chambers is heated and expelled through hydrothermal vents. The undersea geysers build forests of exotic chimneys, called black smokers, that spew out hot water blackened with sulfur compounds.

The black smokers are hosts to the Earth's most bizarre biology (Fig. 89). Flourishing among the hydrothermal vents are perhaps the strangest animals ever encountered due to their unusual habitat. Large white clams up to a foot long

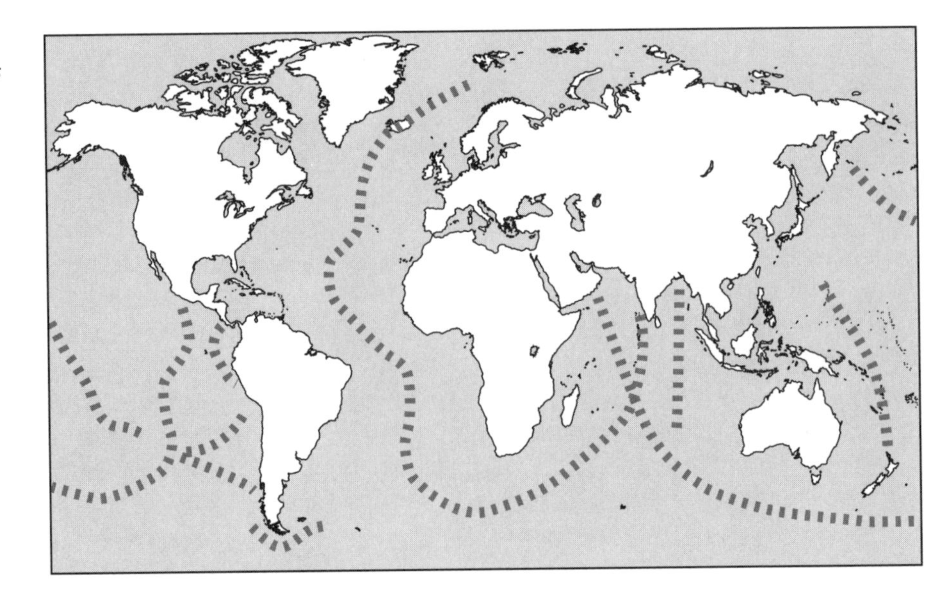

Figure 88 *Midocean ridges, where crustal plates are spreading apart, comprise the most extensive mountain chains in the world and are centers of intense volcanic activity.*

Figure 89 *Cluster of tube worms and sulfide deposits around hydrothermal vents near the Juan de Fuca Ridge.*
(Photo courtesy USGS)

nestle between black pillow lava. Giant white crabs scamper blindly across the volcanic terrain. Clusters of giant tube worms up to 10 feet tall sway in the ocean currents. They obtain their nutrition from bacteria that metabolize sulfur compounds in the hydrothermal water, making their world totally independent of the sun for its energy, which comes instead from the Earth's interior.

Off the coast of Washington state, a field of seafloor geysers expels extremely hot brine, at temperatures between 350 and 400 degrees Celsius, into the near-freezing ocean water. Massive undersea volcanic eruptions from fissures on the ocean floor at spreading centers along the East Pacific Rise create large megaplumes of hot water. The megaplumes are produced by short periods of intense volcanic activity and measure up to 50 to 60 miles wide.

Apparently the ridge splits open and spills out hot water while the lava erupts in an act of catastrophic seafloor spreading. In a matter of a few hours, or at most a few days, up to 100 million tons of superheated water gushes from a large crack in the ocean crust several miles long. When the seafloor splits open in such a manner, it releases vast quantities of hot water held under great pressure beneath the surface. The release of massive amounts of hot water beneath the sea might explain why the ocean remains so salty.

DEEP-SEA TRENCHES

The deepest spot in the world is the Pacific Mariana Trench (Table 11), which forms a long line northward from the island of Guam and reaches a depth of nearly 7 miles below sea level. The deep-sea trenches lie off continental margins and island arcs. They are regions of intense volcanic activity, producing the most explosive volcanoes on Earth. Volcanic island arcs fringe the trenches, and each has similar curves and similar volcanic origins. The trenches form in an arc because this is the geometric figure that occurs when a plane, such as a rigid lithospheric plate, cuts or subducts into a sphere, such as the mantle. The trenches are also sites of almost continuous earthquake activity deep in the bowels of the Earth.

As the Pacific plate inches northwestward, its leading edge dives into the mantle, forming the deepest trenches in the world. When a plate extends away from its place of origin at a midocean spreading center, it becomes thicker and colder, as more material from the asthenosphere adheres to its underside in a process known as underplating. Eventually, the plate becomes so dense it loses buoyancy and can no longer remain on the surface. This causes it to sink into the mantle, and the line of subduction creates a deep-sea trench. As the subducted portion of the plate dives into the mantle, the rest of the plate, which might be carrying a continent on its back, is pulled along with it. This is the main driving mechanism for the drifting of the continents.

TABLE 11 THE WORLD'S OCEAN TRENCHES

Trench	Depth (Miles)	Width (Miles)	Length (Miles)
Peru–Chile	5.0	62	3,700
Java	4.7	50	2,800
Aleutian	4.8	31	2,300
Middle America	4.2	25	1,700
Marianas	6.8	43	1,600
Kuril-Kamchatka	6.5	74	1,400
Puerto Rico	5.2	74	960
South Sandwich	5.2	56	900
Philippines	6.5	37	870
Tonga	6.7	34	870
Japan	5.2	62	500

The deep ocean trenches created by descending plates accumulate large amounts of sediment derived from the adjacent continents. The continental shelf and slope contain thick deposits of sediment washed off the continents. When the sediments and their content of seawater are caught between a subducting oceanic plate and an overriding continental plate, they are subjected to strong deformation, shearing, heating, and metamorphism. The sediments are carried deep into the mantle, where they melt to become the source of new molten magma for volcanoes that fringe the deep-sea trenches (Fig. 90).

The seaward boundaries of subduction zones are marked by deep trenches usually found at the edges of continents or along volcanic island arcs. Behind each island arc lies a marginal or back-arc basin, which is a depression in the ocean crust caused by plate subduction. Deep subduction zones such as the Mariana Trench form back-arc basins, whereas shallow ones such as the Chilean subduction zone off South America do not. One classic back-arc basin is the Sea of Japan between China and the Japanese archipelago, which eventually will be plastered against Asia.

RIVER VALLEYS

A river valley is a linear, low-lying track of land traversed by a river or stream and bordered on both sides by higher ground called a floodplain. A narrow valley is not much wider than the river channel itself, whereas a broad valley

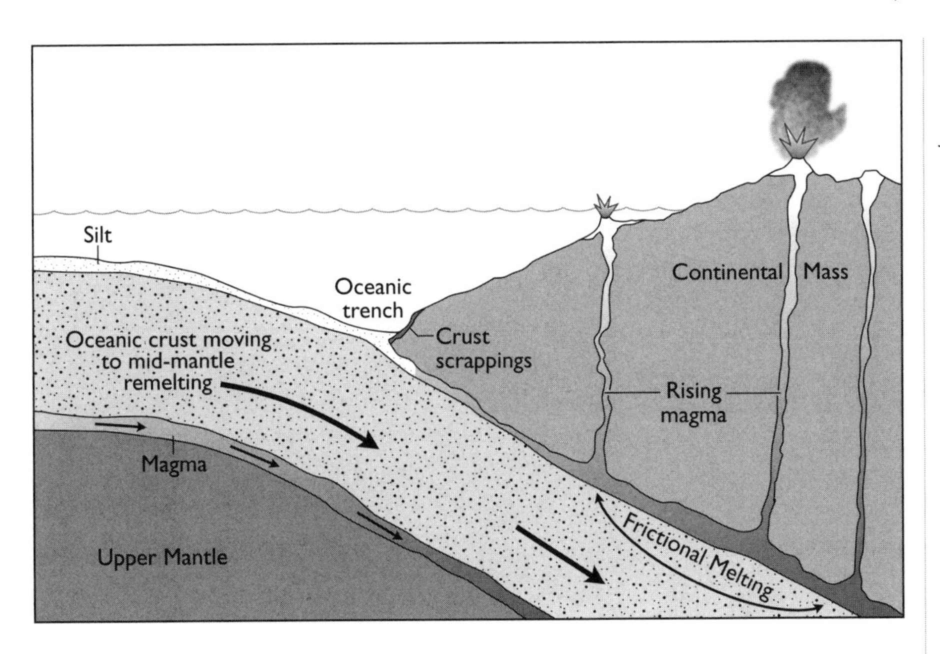

Figure 90 The subduction of the ocean floor provides new molten magma for volcanoes that fringe the deep-sea trenches.

exceeds many times the width of the river channel. A narrow valley is carved out by a fast-flowing river that is actively downcutting in areas of regional uplift, called the youthful stage. Some narrow valleys reside in resistant rocks that slow the lateral cutting of a river and commonly display rapids and waterfalls.

A valley becomes wider as a river flows along a leveling grade and is no longer rapidly downcutting, called the mature stage. This condition occurs mostly near the mouth of a river, where wide floodplains exist. Meanders are common features of wide valleys (Fig. 91), especially in areas with uniform banks composed of easily erodible sediment. The valley might widen by flooding, weathering, and mass wasting. Many river valleys were also widened by glaciers during the ice age, converting V-shaped valleys into U-shaped ones.

Rivers clogged with sediment fill their channels and spill over onto the adjacent plain and carve out a new river course. In the process, they meander downstream, forming thick sediment deposits in broad floodplains that can fill an entire valley. As a stream meanders across a floodplain, the greatest erosion occurs on the outside of the bends, resulting in a steep cutbank in the channel, whereas on the inside of the bends the water slows and deposits its suspended sediments. During a flood, a winding river often takes a short cut across a low-lying area separating two bends, temporarily straightening the river until it further fills its channel with sediment, causing it to meander again. Meanwhile, the cutoff sections of the original river bends become oxbow lakes.

Rivers erode by abrasion and solution. Abrasion occurs when the transported material scours the sides and bottom of the channel. A river transports an enormous amount of debris eroded at its headwaters and along its banks. The sediment is derived from rocks weathered by rain, wind, and ice. A river continues to erode its bed and build it back up. Erosion and sedimentation therefore determine the shape of a river course from one confined to a single straight channel when eroding to one that is meandering or braided when clogged with debris. The sediment-laden stream eventually empties into standing bodies of water (Fig. 92).

River erosion deepens, lengthens, and widens valleys. At the head of a stream, where the slope is steep and water flow is fast, downcutting lengthens the valley by a process called headward erosion, which is mainly how streams cut into the landscape. Farther downstream, both the velocity and discharge increase, while the sediment size and the number of banks decrease, allowing the river to transport a larger load with lesser slope. Bends in the stream slow the river flow by lowering the gradient.

Erosion widens a stream valley by creep, landsliding, and lateral cutting. The process is most pronounced on the outsides of irregular curves, where the valley

Figure 91 *View northwestward across Long Valley showing the meandering North Fork of the Payette River, Valley County, Idaho.*

(Photo by D. L. Schmidt, courtesy USGS)

Figure 92 *The Yahtse River delta, Icy Bay, Alaska.*

(Photo by J. H. Hartshorn, courtesy USGS)

side might be undercut by flowing water. Therefore, streams with migrating curves tend to widen their valleys. Many streams have distinctive symmetrical curves called meanders that distribute the river's energy uniformly.

If a river captures a nearby stream, called "piracy," it creates a larger expanse of flowing water. The river grows at the expense of other streams and becomes dominant because it contains more water, erodes softer rocks, or descends a steeper slope. The river therefore has a faster headward erosion that undercuts the divide separating it from another stream, and captures its water.

DESICCATED BASINS

The Mediterranean Sea is nearly a completely enclosed basin, whose abyssal plains lie more than 10,000 feet deep. The basin holds nearly a million cubic miles of seawater. The evaporation rate is extremely high, and nearly 5 feet of the water's surface evaporates every year, amounting to approximately 1,000 cubic miles of seawater. Only about 10 percent of the loss is compensated by the influx of freshwater from rivers. The rest must be made up by seawater flowing in from the Atlantic Ocean through the narrow Strait of Gibraltar. The high salinity content makes the water heavier than normal seawater, causing it to sink to the bottom. Eventually, highly saline water will fill the entire basin.

Some 6 million years ago, the Mediterranean Basin was completely cut off from the Atlantic Ocean when an isthmus was created at Gibraltar by the northward movement of the African plate, forming a dam across the strait. Over a period of about 1,000 years, the entire sea evaporated. On the bottom, salt deposits formed as the salt content of the water column precipitated and was deposited on the basin floor. Subsequent sedimentation buried trillions upon trillions of tons of salt beneath younger sediments. The dry basin was more than a mile below the continental shelf and probably looked like a much larger version of today's Death Valley.

On the floor of the Mediterranean Sea is an array of salt domes several miles in diameter and up to thousands of feet high. They formed when salt diapirs buried below the seabed were forced toward the surface. Their presence indicates that extensive salt deposits underlie the floor of the Mediterranean. The salt deposits are interbedded with windblown sediments, indicating the former presence of a dry basin. The evaporite deposits are about a mile thick and formed when the entire water column completely evaporated several times over a period of about a million years.

Rivers draining into the desiccated basin gouged out deep canyons. A deep sediment-filled gorge follows the course of the Rhône River in southern France for more than 100 miles and extends to a depth of 3,000 feet below the surface where the river drains into the Mediterranean Sea. Buried under the sediments of the Nile Delta (Fig. 93) is a 1-mile-deep canyon comparable in size to the Grand Canyon. After about a million years, Gibraltar subsided, and the dam was broken. This created a spectacular waterfall, disgorging water at a rate of 10,000 cubic miles a year. The waterfall would have been 100 times larger than the mile-wide Victoria Falls of southern Africa, one of the greatest falls on Earth, and 50 times higher than the Niagara Falls of North America.

Several centuries were required to refill the basin. The refilling of the Mediterranean Sea lowered the global sea level by as much as 35 feet. The mass of the water pressing down on the basin was comparable to the weight of the great ice sheet that spread across Europe during the last ice age.

The Black Sea might have had a similar fate. As with the Mediterranean, it is a remnant of an ancient equatorial ocean called the Tethys Sea that separated Africa from Europe and connected the Atlantic with the Indian Ocean. A collision of the African plate with Europe and Asia 20 million years ago squeezed out the Tethys, resulting in a long chain of mountains and two inland seas. One was the ancestral Mediterranean, and the other was a composite of the Black, Caspian, and Aral Seas, called the Paratethys Sea, which covered much of Eastern Europe. About 15 million years ago, the Mediterranean and the Paratethys separated, and the Paratethys became a brackish sea, much like the Black Sea of today. The disintegration of the great inland waterway was

Figure 93 *The Nile River Delta, viewed from the Space Shuttle.*
(Photo courtesy NASA)

closely associated with the sudden drying of the Mediterranean. In a brief moment (geologically speaking), the Black Sea became almost dry. Then during the last ice age, it refilled again and became a freshwater lake.

About 7,500 years ago, the Black Sea was inundated by a giant flood. An ancient coastline deep below the surface outlined what was once a much smaller freshwater lake as indicated by the discovery of fossils of freshwater mollusks buried in the sediments. Apparently, when the European glaciers melted, the Mediterranean Sea gushed through the Bosporus channel into the Black Sea, until then a landlocked lake. The flood raised water levels half a foot a day. People living near shore would have witnessed the sea advancing on them by as much as a mile daily. By the time the flood ended, the sea was up about 500 feet, and some 60,000 square miles, an area the size of Florida, was underwater. Perhaps this rapid inundation of the sea led to many legends of a great flood, including the biblical story of Noah.

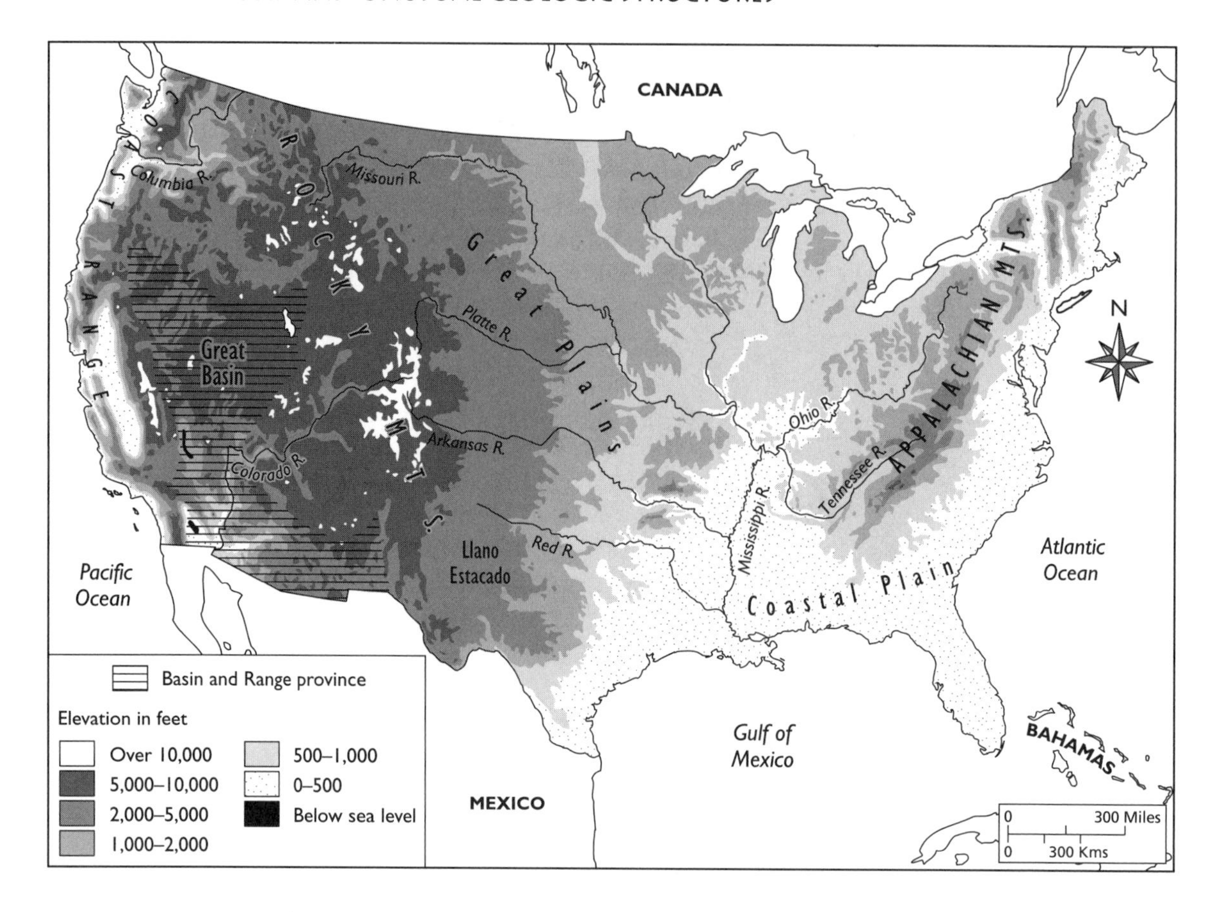

Figure 94
Physiography of the United States.

THE GREAT BASIN

The Basin and Range Province (Fig. 94) is a 600-mile-wide region that includes the Great Basin area of Nevada and Utah. It covers the states of southern Oregon, Nevada, western Utah, southeastern California, and southern Arizona and New Mexico. The Great Basin is a 300-mile-wide closed depression formed by the stretching and thinning of the crust. The crust in the entire Basin and Range Province is actively spreading apart because of forces originating in the Earth's mantle. These are the same forces that raised the Rocky Mountains. The dominant element of the deformation is extension along a line running roughly northwest to southeast.

As the crust continues to spread apart, some blocks sink to form grabens, which are fault-bounded valleys. Between the grabens are ridges called horsts, which are uplifted fault blocks. About 20 horst and graben structures extend from the 2-mile-high scarp of the Sierra Nevada in California to the still-rising Wasatch Front, a major fault system running roughly north-south through Salt Lake City, Utah.

The horst-and-graben structures trend nearly perpendicular to the movement of blocks of crust as they spread apart. This movement is part of the so-called "missing motion" between the Pacific and the North American plates. The Pacific plate is sliding northwestward past the North American plate at about 2 inches per year. The relative motion of these two enormous sections of the Earth's crust is responsible for the deformation of the crust and the ongoing tectonic activity taking place across the western United States. This includes the expansion of the Basin and Range Province, the horizontal slippage along the San Andreas Fault of California, and the uplifting of California's Coast Ranges.

The San Andreas Fault absorbs between 60 and 80 percent of the relative motion between the Pacific and North American plates. The deformation of California's rugged Coast Ranges takes up about 10 percent of the motion, causing intense folding and thrust faulting. The Garlock Fault in California (Fig. 95) is a major east-trending fault, whose left-lateral movement combined with the right-lateral movement of the San Andreas is causing the Mojave Desert to move eastward with respect to the rest of California. The faults of the Mojave and adjacent Death Valley have absorbed between 10 and 30 percent of the total slippage between the Pacific and North American plates over the last several million years.

About 20 million years ago, the Sierra Nevada and the Cascade Ranges were some 170 miles northwest of their present locations. Over the past 20 million years, the Sierra Nevada and the Cascades have swung southwest, opening somewhat like a door with its hinges in the Pacific Northwest. As

Figure 95 *The Garlock Fault in the El Paso Mountains, Bernardino County, California.*

(Photo courtesy USGS)

these ranges moved, the crust to the east was stretched until it ruptured along faults. This stretching and faulting eventually formed the characteristic mountains and valleys of the Basin and Range Province.

The peaks of the Basin and Range formed as the crust was uplifted and stretched and pulled apart, causing huge blocks of crust to slip past one another along faults, rotating and tilting as they slipped. Mountains formed at the tops of the blocks and V-shaped valleys formed at the base. The floors of the valleys eventually flattened as they filled with sediment washed down from the mountains. Continued crustal stretching and faulting in this region could eventually lead to major rifting, creating a rift valley in the desert Southwest similar to the East African Rift Valley that is tearing that continent apart.

Many parallel faults slice through the Basin and Range Province. They absorb approximately 20 percent of the motion between the Pacific plate and the relatively stable North American continent east of the Great Basin. The Basin and Range Province contains fault block mountain ranges bounded by high-angle normal faults. About 5 million years ago, crustal stretching tore Nevada at the heart of the Basin and Range into fault block mountains ranges. The stretching created some of the thinnest crust on any continent in the world. The crust in Nevada is between 12 and 20 miles thick, about half as thick as most continental crust.

The stretching and faulting gradually moved westward, peaking in the Death Valley region about 3 million years ago. This process transformed a land of once gently curving hills into the stark mountains and basins present today. The crust in this region is broken into hundreds of segments that were steeply tilted and raised nearly a mile above the basin, forming nearly parallel mountain ranges up to 50 miles long. The rocks that make up these ranges are greatly deformed by folding, faulting, and igneous and metamorphic activity.

The region is literally being stretched apart due to the weakening of the crust by a series of downdropped blocks. Where the crust was weakened from the expansion, magma welled up to the surface, resulting in massive volcanic eruptions. Volcanic rocks extend into the northwest corner of the Basin and Range Province and have been deformed by faulting. Thrust faults with many miles of displacement run from range to range. Between the ranges are basins, where dry lake bed deposits are common because many low-lying areas once contained lakes. Great Salt Lake and the Bonneville Salt Flats in Utah are good examples.

The formation of the Great Divide Basin of south-central Wyoming occurred under similar circumstances as the Great Basin. It is a topographic and structural basin that covers about 2,500 square miles. The basin is an area of low relief, with elevations ranging from about 6,500 to 7,500 feet. The drainage in the basin is interior with no outlets, and streams and lakes in the region are generally intermittent. A major episode of folding and faulting

Figure 96 *A view northeast across Death Valley to the Black and Funeral Mountains, Inyo County, California.*

(Photo by W.B. Hamilton, courtesy USGS)

occurred during the Laramie orogeny (mountain building episode), which raised the Rocky Mountains and was responsible for the basin's subsidence.

Death Valley, California (Fig. 96), is the lowest spot on the North American continent. Although presently 280 feet below sea level, it had once been at an elevation of several thousand feet. The area collapsed when the continental crust thinned due to extensive block faulting in the region. The Great Basin area is a remnant of a broad belt of mountains and high plateaus that collapsed in a similar manner after the crust was pulled apart.

After discussing some of the great depressions in the Earth's crust, the next chapter will examine the arid and coastal regions.

7

DESERTS AND SEACOASTS

WINDBLOWN AND BEACH SANDS

This chapter examines the geologic features that define the arid and coastal regions. Deserts are among the most dynamic landscapes, constantly changing by drifting sands. Gigantic dust storms and sandstorms prevalent in desert regions play a major role in shaping the arid landscape. Powerful sandstorms clog the skies with thousands of tons of sediment. Roving sand dunes driven across the desert by strong winds engulf everything in their paths. Coastal deserts are unique because they are areas where the seas meet the desert sands. The Namib Desert (Fig. 97) along the coast of Namibia, Africa, is perhaps the world's largest coastal desert.

The Earth is a constantly evolving planet, with complex activities such as running water and moving waves. Rivers carry to the sea a heavy load of sediments washed off the continents, continually building up the coastal regions. Seacoasts vary dramatically in topography, climate, and vegetation. They are places where continental and oceanic processes converge to produce a landscape that is invariably changing on a rapid scale.

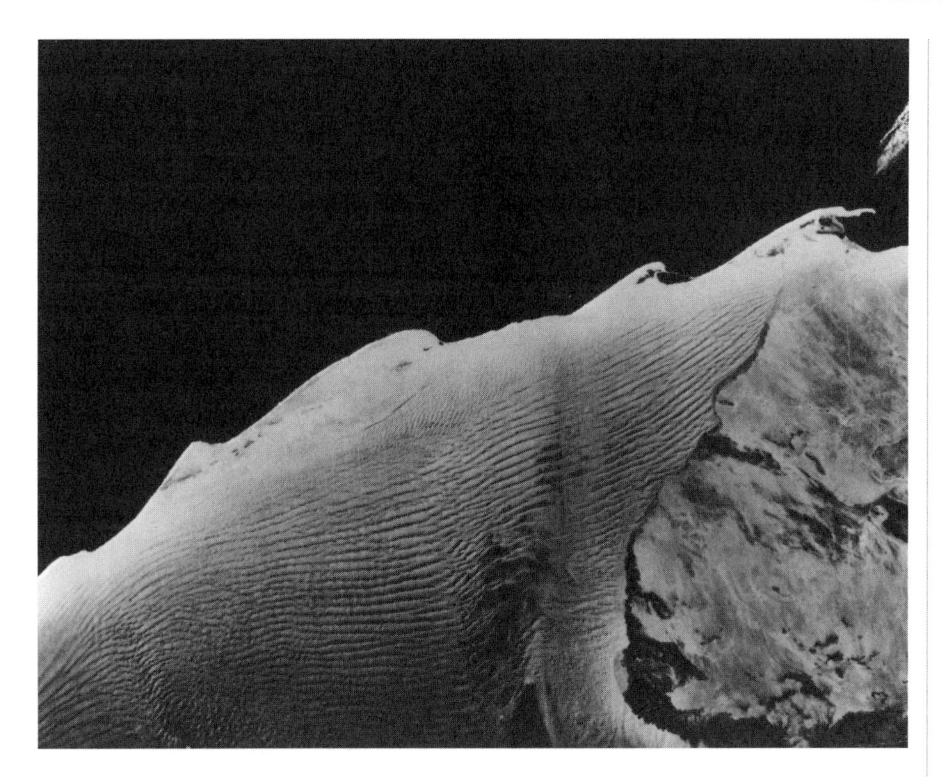

Figure 97 *Linear dunes in the northern part of the Namib Desert, southwest Africa.*

(Photo by E. D. McKee, courtesy USGS)

DESERT FEATURES

The world's deserts are more than just barren landscapes mostly lacking vegetation and animal life. They are among the most dynamic landforms, constantly changing by moving sands. Sometimes, sand dunes cover over human settlements and other constructions, often causing considerable damage. Sandstorms are particularly hazardous, as thousands of tons of sediment clog the skies and land in places where it is unwanted.

About a third of the Earth's landmass is desert. The arid lands are the hottest and driest regions and among the most barren environments. Desert wastelands receive only minor precipitation during certain seasons, with some areas having been essentially rainless for years. Only the hardiest of plant and animal species, some with very unusual adaptations, can tolerate these arid conditions. Often, when the rains arrive, heavy downpours cause severe flash floods that sweep away massive quantities of sediment and debris. Sand dunes driven across the desert by strong winds engulf everything in their paths.

Most of the world's deserts lie in the subtropics in a broad band between 15 and 40 degrees latitude on both sides of the equator. High precipitation levels in the Tropics leave little moisture for the subtropics, where the dry air cools and sinks, producing zones of semipermanent high pressure called blocking highs because they block advancing weather systems from entering the region. Mountains also block weather systems by forcing rain clouds to rise and precipitate on the windward side of the range. The lack of precipitation on the leeward side or opposite end of the mountains results in a rain deficit, creating deserts such as those in the southwestern United States. Moist winds from the Pacific Ocean cool and precipitate as they rise over the Sierra Nevada and other mountain ranges in California, leaving regions to the east parched and dry.

Deserts exhibit a variety of unique geologic features. A dry lake bed called the Racetrack Playa (Fig. 98) at the north end of Death Valley becomes a shallow lake during heavy rains. The area is known for its mysterious roving rocks, which leave tracks in the lake bed that have puzzled geologists for many decades. Originally, high winds whistling off the nearby mountains were thought to be powerful enough to push the boulders through the muddy lake bed after a soaking rain. However, the largest of these boulders measured 2 feet across and weighed some 700 pounds, much too heavy for the wind to push around.

If, instead, a thin layer of ice formed on the 3,700-foot-altitude lake bed after a winter rain, it could lift the rocks slightly, reducing the frictional contact between the boulder and the mud. With the aid of a strong wind, the inch-thick sheet of ice would skim across the lake bed with embedded rocks etching patterns in the mud. Tracks as much as 2,500 feet apart made identi-

Figure 98 *The Racetrack showing mud-cracked playa and pebble-size playa scrapers that dug furrows, Death Valley National Monument, Inyo County, California.* (Photo by W. B. Hamilton, courtesy USGS)

Figure 99 *Alluvial fans from the Sierra Nevada, Death Valley National Monument.*

(Photo by H. E. Malde, courtesy USGS)

cal patterns, indicating the track-making rocks were locked into the same sheet of ice. Often, when the ice breaks apart, the rocks go off on their own, making curlicues and wiggles in the mud.

Further examples of patterned ground include polygonal shapes created in Death Valley's desert muds. The mud cracks formed when mud contracted as it rapidly dried in the hot sun. Sorted circles were also created by the increased wearing down of coarse grains in isolated cracks in bedrock. Even earthquake vibrations are believed to cause the sorting of some sediments.

Erosion of mountain ranges in desert regions relies on heavy downpours along small drainage areas. Sediment fans consisting of sands and gravels develop at the mountain front. When the formerly steep mountain front retreats, it leaves a smooth surface in the bedrock called a pediment, which generally has a concave-upward slope of up to 7 degrees, depending on the sediment size and the amount of runoff. Streams issuing from the mountains change course back and forth across the pediment, forming alluvial fans (Fig. 99). Eventually, the mountains erode down to the level of the plain, leaving the pediments speckled with remnants of the range.

Development of drainage patterns in desert lands is well demonstrated in the Basin and Range Province of the American Southwest. The region con-

Figure 100 *The south end of the Great Salt Lake Desert, Utah, showing alkali flats.*

(Photo by C. D. Walcott, courtesy USGS).

tains several mountain ranges formed by relatively recent faulting. The basins between ranges are low-lying areas that often contain lakes during wetter climates. Lake-deposited sediments are common, and dry lake beds called playas cover the surface. The bodies of water are called alkali lakes because of their high concentrations of salt and other soluble minerals. When the lakes evaporate, they become alkali flats (Fig. 100).

The Dead Sea in the Syrian Desert on the border between Israel and Jordan at 1,300 feet below sea level is the lowest place on the continents. It occupies the Jordan Rift zone, a deep gash in the crust where the land is being pulled apart. It is also one of the world's deepest lakes, some 1,000 feet in depth. For thousands of years, rivers laden with salts leached from the rocks have flowed south through the Jordan Rift Valley and terminated in the Dead Sea. With no outlet, the inflowing water evaporates into the dry desert air, which concentrates the salts, making the Dead Sea the world's saltiest lake with an average salinity eight times higher than the ocean.

The Sahara Desert (Fig. 101) is the largest arid region on Earth, with an area of about 3.5 million square miles, or about the size of the United States. Deep beneath its sands lies a vast network of ancient river valleys and smaller stream channels that wind across the bedrock, including gravel terraces, desert basins, and other geologic structures. A search through the sands uncovered one of the last great river systems in the world as wide as Egypt's Nile Valley. The discovery suggests that buried riverbeds exist in other deserts as well.

An interesting drainage system was found buried beneath the sands of the eastern Sahara Desert. It was one of the last great river systems on Earth,

Figure 101 *The Sahara Desert of central Algeria showing linear dunes.*

(Photo by E. D. McKee, courtesy USGS)

whose activity was much more intense than had been previously thought. Channels hundreds of thousands of years old twisted through valleys that are millions of years old but are now filled with thick deposits of sand. Buried valleys lying under the sand might once have been migration routes for early humans leaving Africa for Europe and Asia. Scattered in the sand are Stone Age human artifacts, suggesting that humans and life-sustaining water sources were once present in an area that is now utterly uninhabitable. Dozens of human artifacts, including stone axes up to a quarter million years old, appeared to mark ancient campsites, where people called *Homo erectus* lived and made stone tools.

Geologists exploring for petroleum in the Sahara stumbled upon a series of giant grooves cut into the underlying strata apparently by glaciers during an ancient ice age. Rocks embedded at the base of the glacial ice scoured the landscape as the massive ice sheets moved back and forth. Evidence suggesting that thick sheets of ice once blanketed the region included erratic boulders dumped in heaps by the glaciers along with eskers, which are long sinuous sand deposits from glacial outwash streams.

Possibly the most impoverished deserts on Earth are the dry valleys running between McMurdo Sound and the Transantarctic Range in Antarctica. Because they are protected by the mountains against snowstorms, the dry valleys receive less than 4 inches of snow annually, most of which blows away by hurricane-force winds reaching 200 miles per hour and more. Some areas have not received any form of precipitation for up to a million years.

WIND EROSION

In desert regions, wind is the most active agent of erosion, transportation, and deposition. Wind plays an important part in distributing sand, by setting particles in motion. They tend to bump and wiggle along, sometimes blown a foot or more above the ground. Wind erosion occurs mainly by deflation, which is the removal of large amounts of sediment by windstorms, resulting in a deflation basin. In some areas, deflation produces hollows called blowouts (Fig. 102), which are recognized by their typically concave shape. Often after the fine material has been removed, a layer of pebbles remains to protect against further deflation.

The deserts are host to some of the strongest winds due to the rapid heating and cooling of the land surface. The winds generate sandstorms and dust storms, which work together to cause wind erosion. Dust storms often involve a solid wall of dust blowing at speeds of 60 miles per hour or more. The rising dust clouds can extend several thousand feet high and stretch for hundreds of miles.

Figure 102 *A blowout with core 3 miles south of Harrison, Sioux County, Nebraska.*

(Photo by N. H. Darton, courtesy USGS)

The land is scoured by the winds, and several inches of soil can be airlifted to other areas. Massive dust storms called haboobs, from Arabic meaning "violent wind," arise in the deserts of Africa, Arabia, central Asia, Australia, and the American Southwest (Fig. 103 and Table 12). Enormous dust storms occur when a powerful airstream moves across vast deserts such as those in Africa, where huge dust bands up to 1,500 miles long and 400 miles wide race across the desert floor.

Some large African storm systems have been known to carry dust all the way across the Atlantic Ocean to South America. The dust over African deserts rises to high altitudes, where westward-flowing air currents transport it to the Americas. Fast-moving storm systems in the Amazon rain forest pull in the dust, which contains nutrients that enrich the soil. So much African dust blows across the Atlantic during summer storms that Florida and other areas along the East Coast of the United States actually violate clean air standards.

The Sahara dust blows across much of the United States, possibly reaching as far as the Grand Canyon, contributing to the notorious haze that obscures the canyon's beauty. The dust is chemically different from local soils and has a distinctive red-brown color. When added to other air pollutants, the Sahara dust causes a persistent haze, especially in summer. The dust has an unexpected benefit, however. The periodic influxes help regions plagued by acid rain produced by burning fossil fuels by diluting the acidic continent of

Figure 103 *A massive dust storm in the western plains of the United States.*

(Photo courtesy USDA-Soil Conservation Service)

TABLE 12 MAJOR DESERTS

Desert	Location	Type	Area (Square Miles X 1,000)
Sahara	North Africa	Tropical	3,500
Australian	Western/interior	Tropical	1,300
Arabian	Arabian Peninsula	Tropical	1,000
Turkestan	Central Asia	Continental	750
North America	S.W. U.S./N. Mexico	Continental	500
Patagonian	Argentina	Continental	260
Thar	India/Pakistan	Tropical	230
Kalahari	S.W. Africa	Littoral	220
Gobi	Mongolia/China	Continental	200
Takla Makan	Sinkiang, China	Continental	200
Iranian	Iran/Afghanistan	Tropical	150
Atacama	Peru/Chile	Littoral	140

rainwater. Coral off the Florida Keys trap the dust inside growth bands, which can be used to trace dust from sources, such as storms of sand blowing off the Sahara Desert toward the United States.

In dry regions where dust storms were once prevalent, the wind transported large quantities of loose sediment. Most windblown sediments accumulated into thick deposits of loess (Fig. 104), which is a fine-grained, loosely consolidated, sheetlike deposit that often shows thin, uniform bedding on outcrop. Loess deposits cover thousands of square miles. Secondary loess deposits were transported and reworked over a short distance by water or intensely weathered in place.

Loess sediments are common in North America, Europe, and Asia. China contains the world's largest deposits, which originated from the Gobi Desert and attain hundreds of feet in thickness. The sediment comprises angular particles of equal grain-size composed of quartz, feldspar, hornblende, mica, and bits of clay. It is usually a buff to yellowish brown loamy deposit that is commonly unstratified due to a rather uniform grain size, generally in the silt-size range. Loess often contains the remains of grass roots, and as with mud bricks, deposits can stand in nearly vertical walls despite their weak cohesion. Loess also can cause problems in construction unless properly compacted because on wetting it tends to settle.

Over thousands of years, deserts developed a protective shield of pebbles coated with a desert varnish, ranging in size from a pea to a walnut and are too heavy for the strongest desert winds to pick up. Military activities during the 1991 Persian Gulf War have disturbed extensive areas of the desert in Kuwait, northeast Saudi Arabia, and southern Iraq, where the natural desert

Figure 104 *An exposure of loess standing in vertical cliffs, Warren County, Mississippi.*

(Photo by E. W. Shaw, courtesy USGS)

shield has been breached. The desert shield helps hold down sand and dust particles and creates a stable terrain. Without this cover, the desert could spawn a new generation of sand dunes and a higher incidence of dust storms that sweep the sands from place to place.

Wind erosion results in deflation and abrasion. Deflation is the removal of sand and dust particles by the wind, which often excavates hollowed-out areas. It usually occurs in arid regions and unvegetated areas such as deserts and dry lake beds. As smaller soil particles blow away during dust storms, the ground coarsens over time. The remaining sand tends to roll, creep, or bounce with the wind until it meets an obstacle, whereupon it settles and builds into a dune.

The process of abrasion is similar to sandblasting by wind-driven sand grains that can cause erosion near the base of a cliff. When acting on boulders or pebbles, abrasion pits, etches, grooves, and scours exposed rock surfaces (Fig. 105). Abrasion also produces some unusually-shaped rocks called ven-

Figure 105 *A wind blowout in Fremont County, Idaho in August 1921.*

(Photo by H. T. Stearns, courtesy USGS)

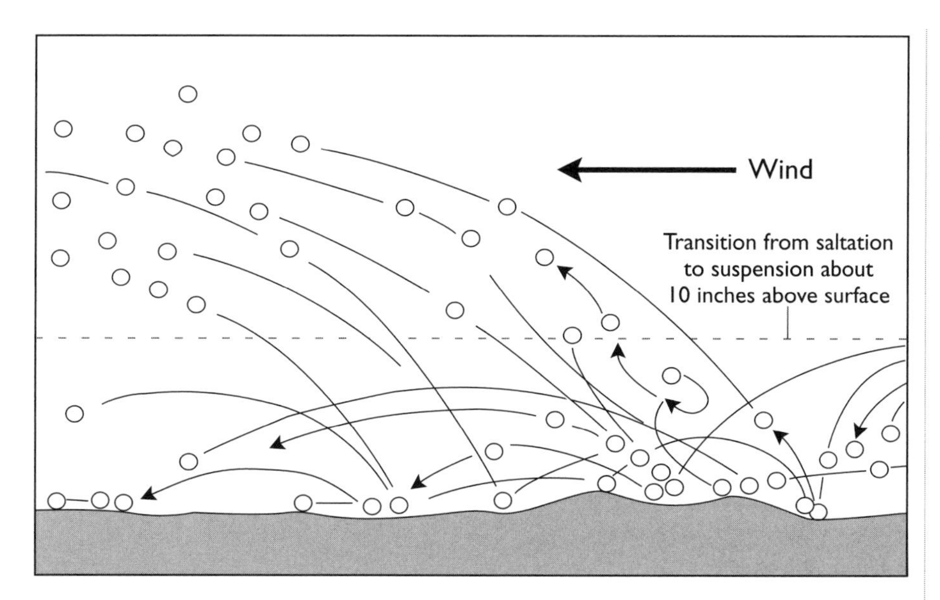

Figure 106 *The birth of a dust storm during the transition of sand grains from saltation to suspension.*

tifacts, which often have several flat, polished surfaces depending on the wind direction or the movement of the rock.

Abrasion by sand blasting erodes the surface of boulders, while desert varnish, composed of iron and magnesium oxides exuded from the rock, colors them dark brown or black. Maximum erosion effects occur during strong sandstorms, with sediment grains generally rising less than 2 feet above the ground. These abrasive effects are most commonly seen on fence posts and power poles.

Sands behave in mysterious ways, acting partly as solids and partly as liquids. Sand grains march across the desert floor under the influence of strong winds by a process called saltation (Fig. 106). The grains of sand become airborne for an instant, rising a foot or more above the ground. When landing, they dislodge additional sand grains, which repeat the process. The rest of the moving sand travels forward along the surface by rolling and sliding. The sediment grains of desert deposits are often frosted due to abrasion from the constant motion of the sand.

SAND DUNES

Sand dunes driven across the desert by powerful wind currents comprise about 10 percent of the world's arid regions. The dunes move across the desert floor in response to the wind as sand grains in motion dislodge one another and become airborne for a moment. The size and shape of sand dunes are determined by the direction, strength, and variability of the wind, the soil moisture content, the vegetative cover, the underlying topography, and the quantity of movable soil exposed to the wind.

Sand dunes marching across the desert floor engulf everything in their paths, causing major problems in the construction and maintenance of highways and railroads that cross sandy areas of deserts. Sand dune migration near desert oases creates another serious problem, especially when encroaching on villages. Damage to structures from sand dunes can be reduced by building windbreaks and by funneling sand out of the way. Without such measures, disruption of roads, airports, agricultural settlements, and towns could pose many difficulties for desert regions.

The three basic shapes of sand dunes is determined by the topography of the land and patterns of wind flow. Linear dunes (see Fig. 101) are aligned in roughly the direction of strong steady prevailing winds. They are significantly longer than they are wide and parallel each other, sometimes producing a wavy pattern. When the wind blows over the dune peaks, part of the air flow shears off and turns sideways. The air current scoops up sand and deposits it along the length of the dune, which maintains its height and simultaneously lengthens it. The surface area covered by dunes is about equal to the area between dunes. Both sides of the dune are likely steep enough to cause avalanches.

Barchans or crescent dunes are symmetrically shaped with horns pointing downwind. They travel across the desert at speeds of up to 50 feet a year. Parabolic dunes form in areas where sparse vegetation anchors the side arms, while the center blows outward and moves sand in the middle forward. Star or radial dunes (Fig. 107) form by shifting winds that pile sand into central points that can rise 1,500 feet and more. They have several arms projecting outward, resembling giant pinwheels. Sand also accumulates in flat sheets or forms stringers downwind that do not exhibit any appreciable relief in sand seas.

Sand dunes marching across the desert floor exhibit an unexplained phenomenon known as booming sands. The sound occurs almost exclusively in large, isolated dunes deep in the desert or on back beaches well inland from the coast. The noises can be triggered by simply walking along the dune ridges. When sand slides down the lee side of a dune, it sometimes emits a loud rumble. The sounds emitting from the dunes have been likened to bells, trumpets, pipe organs, foghorns, cannon fire, thunder, buzzing telephone wires, or low-flying aircraft. The grains in sound-producing sand are usually spherical, well-rounded, and well-sorted, or of equal size. The sound appears to originate from a harmonic event occurring at the same frequency. However, normal landsliding involves a mass of randomly moving sand grains that collide with a frequency much too high to produce such a peculiar sound. At least 30 "booming" dunes have been found in deserts and on beaches in Africa, Asia, North America, and elsewhere.

Figure 107 *Compound star dunes in Gran Desierto, Sonora, Mexico.*
(Photo by E. D. McKee, courtesy USGS)

BEACH SANDS

Marine sediments comprise quartz grains about the size of beach sands, and indeed many marine sandstone formations such as those exposed in the American West were deposited along the shores of ancient inland seas. Gravels are rare in the ocean and move mainly from the coast to the deep abyssal plains by submarine slides. Windblown sediments landing in the ocean slowly build deposits of abyssal red clay, whose color derived from iron oxide signifies its terrestrial origin.

Rivers and streams deliver to the ocean a heavy load of sediment washed off the continents. When reaching the coast, the riverborne sediments settle out of suspension according to grain size. The coarse-grained sediments deposit near the turbulent shore, while the fine-grained sediments settle in calmer waters farther out to sea. As the shoreline advances seaward from a buildup of coastal sediments or a falling sea level, finer sediments are overlain by progressively coarser ones. As the shoreline recedes landward from a lowering of the land surface or a rising sea level, coarser sediments are overlain by progressively finer ones. The differing sedimentation rates as the sea transgresses and regresses result in a recurring sequence of sands, silts, and muds.

The overlying sedimentary layers pressing down on the lower strata and cementing agents such as calcite and silica transform the sediments into solid rock, providing a geologic column of alternating beds of limestones, shales, siltstones, and sandstones. Abrasion eventually grinds all rocks down to clay-size particles, becoming the most abundant sediments. The small minute particles sink slowly, settling out in calm, deep waters far from shore. Each bedding plane marks where one type of deposit ends and another begins. Thus, thick sandstone beds might be interspersed with thin beds of shale and siltstone, indicating periods when coarse sediments were punctuated by periods when fine sediments were laid down, as the shoreline progressed and receded.

Graded bedding results from the varying of particle size in a sedimentary bed from coarse at the bottom to fine at the top, indicating rapid deposition of sediments of differing sizes by fast-flowing streams emptying into the sea. The largest particles settle out first and are covered by progressively finer material due to the difference in settling rates. Beds also grade laterally, resulting in a horizontal gradation of sediments from coarse to fine. Sedimentary beds also vary in color, which helps identify the type of depositional environment, with gray-colored sediments suggesting a marine origin.

Limestones deposited on the shallow floors of oceans or large lakes are among the most common rocks (Fig. 108) and make up about 10 percent of the land surface. They are composed of calcium carbonate mostly derived from biologic activity as evidenced by abundant fossils of marine life in limestone beds. Chalk is a soft, porous carbonate rock, whose poor consolidation of the strata often results in severe erosion during coastal storms. Dolomite resembles limestone, with the calcium in the original carbonate replaced by magnesium. The chemical reaction can cause a reduction in volume and create void spaces. The Dolomite Alps in northeast Italy are upraised blocks of this mineral deposited on the bottom of an ancient sea.

Riverborne sediments entering the ocean settle onto the continental shelf, which extends up to 100 miles or more and reaches a depth of roughly 600 feet. In most places, the continental shelf is nearly flat-lying with an average slope of about 10 feet per mile, comparable to the slopes of many coastal regions. Indeed, during the ice ages these areas were the coastal regions of the world.

SEA CLIFFS

Breaking waves dissipate energy along the coast and are responsible for generating alongshore currents, which transport sand along the beach. They also cause coastal erosion, a serious problem at many seacoasts where the shoreline is steadily receding. Most high waves and beach erosion occur during coastal storms. Hurricanes with winds of 100 miles per hour and more produce the

Figure 108
Rhythmically interbedded limestone and shaly limestone of the Nasorak Formation, Lisborne District, northern Alaska.
(Photo by M. R. Campbell, courtesy USGS)

most dramatic storm surges, which are responsible for destroying entire beaches. The process of beach erosion is mainly influenced by the strength of beach dunes or sea cliffs, the intensity and frequency of coastal storms, and the exposure of the coast.

The erosion of sea cliffs and dunes that mark the coastline causes the shore to retreat a considerable distance. Steep waves accompanying storms at sea seriously erode sand dunes and sea cliffs (Fig. 109). Waves erode by impact and pressure, by abrasion, and to a lesser extent by solution. Therefore, wave erosion is similar to river erosion. Most beach material originates from wave erosion and river deposition. Wave impact can dislodge and transport large fragments. Waves running up onshore and back to sea move sand and pebbles back and forth, abrading sediments while simultaneously carrying them farther out to sea.

Waves breaking on a coastline develop sea cliffs by undercutting the bedrock (Fig. 110). Coastal slides occur when wave action undercuts a sea cliff, causing it to fall into the ocean. Sea cliff retreat results from marine and nonmarine agents, including wave attack, wind-driven salt spray, and mineral solution. The nonmarine agents responsible for sea cliff erosion include chemical and mechanical processes, surface drainage water, and rainfall. Mechanical erosion processes rely on cycles of freezing and thawing of water in crevasses, forcing apart fractures that further weaken the rock.

Weathering agents break down rocks or cause them to shed their outer layers. Animal trails that weaken soft rock and burrows that intersect cracks in the soil also cause sea cliffs to erode. The sea cliff is further eroded by surface water runoff and wind-driven rain. Excessive rainfall along the coast can lubricate sediments, causing huge blocks to slide into the sea. Water running over the cliff edge and wind-driven rain produce the fluting, or grooves, often exposed on cliff faces.

Figure 109 *The sea cliff at Moss Beach, San Mateo County, California has receded 165 feet since 1866.*

(Photo by K. R. Lajoie, courtesy USGS)

Figure 110 *Quaternary wave-cut bench on jasper of Franciscian Formation, San Luis Obispo County, California.*

(Photo by G. W. Stose, courtesy USGS)

Seeping groundwater from a sea cliff can form indentations on the cliff face, which undermines and weakens the overlying strata. The addition of water also increases pore pressure within sediments, reducing the binding strength that holds the rock layers together. If bedding planes, fractures, or jointing dip seaward, water moving along these areas of weakness can induce rock slides. Such slides have excavated large valleys on the windward parts of the Hawaiian Islands, where powerful springs emerge from porous lava flows.

Direct wave attacks at the base of a sea cliff quarry out weak beds and undercut the cliff, causing the overlying unsupported material to collapse onto the beach (Fig. 111). Waves also work along joint or fault planes to loosen blocks of rock or soil. Furthermore, wind carrying salt spray from breaking waves drives it against the sea cliff. Porous sedimentary rocks absorb the salty water, which evaporates, forming salt crystals whose growth weakens rocks. The surface of the cliff slowly flakes off and falls to the beach below, where the material landing at the base of the sea cliff piles up into a talus cone.

Besides wave erosion, limestone cliffs also erode by chemical processes that dissolve soluble minerals in the rocks. Limestone erosion is common on coral islands in the South Pacific, possibly due to rising sea levels and higher temperatures that kill the coral, and on the limestone coasts of the Mediterranean and Adriatic Seas. Seawater dissolves the lime cement in sediments, forming deep notches in the sea cliffs. Chemical erosion also removes cementing agents, causing the sediment grains to separate.

Figure 111 *Highway 1 at the Devils Slide, San Mateo County, California.*

(Photo by R. D. Brown, courtesy USGS)

COASTAL STRUCTURES

The breaking of a large wave on the coast is a striking example of the sizable amount of energy ocean waves generate. The intertidal zones of rocky weather coasts receive much more energy per unit area from waves than they do from the sun. The waves are created by strong winds generated by distant storms blowing across large areas of the open ocean. Wave energy reflecting off steep beaches or seawalls forms sandbars. When waves approach the shore at an angle to the beach, the wave crests bend by refraction. As waves pass the end of a point of land or the tip of a breakwater, a circular wave pattern generates behind the breakwater. When the refracted waves intersect other incoming waves, they increase the wave height.

Local storms near the coasts provide the strongest waves, especially when superimposed on rising tides. High tides generally exceeding a dozen feet called megatides arise in gulfs and embayments along the coast in many parts of the world. Their height depends on the shapes of bays and estuaries, which channel the tides and increase their amplitude. Many locations with extremely high tides also experience strong tidal currents. Tidal floods are overflows on coastal areas bordering the ocean or an estuary. The coastal lands, including bars, spits, and deltas, are affected by coastal currents and offer similar protection from the sea as floodplains do from rivers. Coastal flooding is primarily caused by high tides, waves from strong winds, storm surges (Fig. 112), tsunamis, or any combination of these. Tidal floods also result from high waves combined with flood runoff from heavy rains that accompany coastal storms.

Flooding can extend far along a coastline. The duration is generally short, depending on the elevation of the tide, which usually rises and falls twice daily. If the tide is in, other factors that produce high waves can raise the maximum level of the prevailing high tide. Tidal waves generated by strong winds superimposed on regular tides produce the greatest tidal floods and the most severe beach erosion. Beach erosion is influenced by the strength of beach dunes or sea cliffs, the intensity and frequency of coastal storms, and the exposure of the coast. Not all shoreline retreat can be blamed on rising sea levels alone, but is also determined by long-term changes in the size and direc-

Figure 112 *Overwash and storm surge penetration near Cape Hatteras, North Carolina in 1984.*

(Photo by R. Dolan, courtesy USGS)

tion of waves striking the coast. The rate of coastal retreat varies with the geography of the shoreline and the prevailing wind and tides.

Prevention of beach erosion is often thwarted because the waves constantly batter and erode defenses built to keep out the sea. As a result, the methods developers use to stabilize the seashores are destroying the very beaches upon which they intend to build. Beach erosion is often aggravated by the structures engineers build to stabilize the shoreline. Jetties and seawalls erected to halt the tides generally increase erosion. Jetties cut off the natural supply of sand to beaches, and seawalls increase erosion by bouncing waves back instead of absorbing their energy. The rebounding waves carry sand out to sea, undermining the beach and destroying the shorefront property the seawall was designed to protect. These structures often hasten the erosion of beach sands in front of the wall. In effect, the seawalls are saving the bluffs at the detriment of the beaches. Barriers placed at the bottom of sea cliffs might deter wave erosion, but do not affect sea spray and other erosional processes. Beaches in front of the seawalls often lose sand during certain seasons, while waves return beach sands at other times.

The melting of the polar ice caps during a sustained warmer climate could substantially raise sea levels and drown coastal regions. At the present rate of melting, the sea could rise a foot or more by the middle of this century, comparable to the melting rate of the continental glaciers at the end of the last ice age. Consequently, beaches and barrier islands inevitably disappear as shorelines move inland (Fig. 113).

For every foot of sea level rise, up to 1,000 feet of seashore would be inundated, depending on the slope of the coastline. A 3-foot rise in sea level could flood about 7,000 square miles of coastal land in the United States. The receding shore would result in the loss of large tracks of coastal land along with shallow barrier islands. Estuaries, where marine species hatch their young, would be destroyed. Low-lying fertile deltas that feed much of the world's population would be inundated by the rising waters. Coastal cities, where half the world's human population resides, would have to move farther inland or build seawalls to keep out the rising waters.

CORAL REEFS

Coral reefs are among the most important land builders, forming chains of islands and altering continental shorelines. About 270,000 square miles of coral reefs are estimated to exist in the world's oceans. Over geologic time, corals and other organisms living on the reefs built massive formations of limestone. A typical reef consists of fine, sandy detritus, stabilized by plants and animals anchored to the surface. The coral's ability to build wave-resistant

Figure 113 *Old stumps and roots exposed by shore erosion at Dewey Beach, Delaware, indicate that this area was once the tree zone.*

(Photo by J. Bister, courtesy USDA Soil Conservation Service)

structures thereby encouraged tropical plant and animal communities to thrive on the reefs, which are thought to house one in every four marine species.

Reefs are restricted to warm, shallow waters in the Indo-Pacific and the western Atlantic. Hundreds of atolls comprising rings of coral islands that enclose a central lagoon dot the Pacific Ocean. They consist of reefs several thousand feet across, many of which formed on ancient volcanic cones that have dropped beneath the waves, with the rate of coral growth evenly matching the rate of subsidence.

The coral rampart, which reaches almost to the water's surface, is the major structural feature of the living reef. It consists of large rounded coral heads and a variety of branching corals. Hundreds of species of encrusting organisms such as barnacles thrive on the coral reef. Smaller, more fragile corals and large communities of green and red calcareous algae live on the coral framework.

The fore reef is seaward of the reef crest, where coral blankets nearly the entire seafloor. In deeper waters, many corals grow in flat, thin sheets to maximize their light-gathering area. In other parts of the reef, the corals form large buttresses separated by narrow sandy channels composed of calcareous debris from dead corals, calcareous algae, and other organisms living on the coral. The channels resemble narrow winding canyons with vertical walls of solid coral. They dissipate wave energy and allow the free flow of sediments, which pre-

Figure 114 *A fringing coral reef in Puerto Rico.*

(Photo by C. A. Kaye, courtesy USGS)

vents the coral from choking on the debris. Below the fore reef is a coral terrace, followed by a sandy slope with isolated coral pinnacles, then another terrace, and finally a nearly vertical drop into the dark abyss.

Fringing reefs (Fig. 114) grow in shallow seas and hug the coastline or are separated from the shore by a narrow stretch of water. Barrier reefs also parallel the coast, but lie farther out to sea. They are much larger and extend for longer distances. The best example is the Great Barrier Reef, a chain of more than 2,500 coral reefs and small islands off the northeastern coast of Australia. It forms an undersea embankment more than 1,200 miles long, up to 90 miles wide, and as much as 400 feet high. The reef is the largest feature built by living organisms and among the great wonders of the world.

After an examination of desert and coastal deposits, the next chapter will discuss landforms created by glacial ice.

8

GLACIAL TERRAIN

STRUCTURES FORMED BY GLACIERS

This chapter examines the different types of landforms resulting from glacial erosion and deposition. Much of the landscape in the northern latitudes owes its unusual topography to massive ice sheets that swept down from the polar regions during the ice ages. Glaciation was so pervasive ice sheets 2 miles or more thick enveloped much of upper North America and Eurasia. In some places, the crust was scraped completely clean of sediments, exposing the granitic basement rock below. In other areas, glacial sediments were deposited in massive heaps when the glaciers melted and retreated to the poles.

The ice age is still with us; we just live in a warmer period of it. Within a few thousand years, the ice sheets will again be on the move, wiping out everything in their paths. In their rampage, the huge glaciers will bulldoze northern cities and shove their wreckage far south. Entire forests will disappear, and the denizens of the woodlands will scamper for warmer climes toward the tropics.

THE ICE CAPS

Continental glaciers are the largest ice sheets. During the last ice age from about 100,000 to 12,000 years ago, ice sheets covered one-third of the land surface. Today only Antarctica and Greenland contain substantial ice masses, with about 30 percent of the total ice volume that existed during the last ice age. A continental glacier moves in all directions outward from its point of origin and completely engulfs the land except isolated high mountain peaks projecting above the surface of the ice. The term *ice cap* also describes a small glacier that spreads out radially from a central point, as on Iceland.

The largest ice sheet overlies Antarctica. The continent has geographical features similar to those of other continents, only its mountain ranges, high plateaus, lowland plains, and canyons are buried under a sheet of ice in places as much as 3 miles thick (Fig. 115). Antarctica is divided by a wall of mountains that forms the spine of the continent called the Transantarctic Range. It separates the eastern and western ice sheets into a large East Antarctic ice mass and a smaller West Antarctic lobe about the size of Greenland. The ice cap averages about 1.3 miles thick with a mean elevation of about 7,500 feet above sea level. Barren mountain peaks soar 17,000 feet above the ice sheet, and hurricane-force winds shriek off the ice-laden mountains and high ice plateaus.

Figure 115 *The Antarctic Peninsula ice plateau showing mountains literally buried by ice.*

(Photo by P. D. Rowley, courtesy USGS)

Figure 116 *The Wright Dry Valley, Taylor Glacier region, Victoria Land, Antarctica.*

(Photo by W. B. Hamilton, courtesy USGS)

Antarctica is literally a desert, with an average annual snowfall of less than 2 feet, which translates into roughly 3 inches of rain, making the continent one of the driest regions. However, because little melting occurs in the continental interior, any new snowpack adds to the growth of the ice sheets. Dry valleys (Fig. 116) gouged out by local ice sheets running between McMurdo Sound and the Transantarctic Mountains are the largest ice-free areas on the continent. They receive less than 4 inches of snowfall yearly, most of which blows away by strong winds. The landscapes of the dry valleys are very old, and some surfaces are quite steep and appear to have remained virtually unchanged for 15 million years.

About 90 percent of the Earth's ice lies atop Antarctica, whose glaciers contain 70 percent of all the world's freshwater. The bulk of the ice has not changed significantly over the past 15 million years, when the Earth took a plunge into a colder climate. Trapped under the thick polar ice cap is a huge expanse of water the size of Lake Ontario, covering more than 5,000 square miles to a depth of at least 1,600 feet. Water collected in bedrock pockets was prevented from freezing by geothermal heat from below and pressure from the ice above.

Sea ice surrounding Antarctica expands to nearly 8 million square miles in winter, more than twice the size of the United States. It grows at an average rate of more than 20 square miles per minute and usually attains a thickness of no more than 3 feet. Antarctic sea ice differs from that in the Arctic, where most of the ocean is surrounded by land, which dampens the seas and allows the ice to grow over twice as thick. Some Arctic ice survives the summer, so in 4 years it doubles in thickness. However, in the Antarctic, powerful storms at sea churn the water and break up the ice, preventing it from growing any thicker.

Floating ice shelves covering the Ross and Weddell Seas dominate West Antarctica. The elevation in this region is generally low, and most of the ice rests on glacial till lying mostly below sea level. The till is a mixture of ground-up rock and water that acts as a lubricant to aid the ice sheet's slide into the ocean. The Filchner-Ronne Ice Shelf south of the Weddell Sea, the most massive floating block of ice on Earth, has two distinct layers. The top layer measures about 500 feet thick and is composed of ice mostly formed by falling snow. The bottom layer measures about 200 feet thick and consists of frozen seawater. The freshwater layer contains opaque and granular ice similar to the upper portion of a glacier. In contrast, the transparent marine shelf ice displays many inclusions of marine origin such as plankton and clay particles. Free-floating ice platelets recrystallize at the base of the marine layer, forming a slush that compacts into solid ice.

The ice sheet on East Antarctica rests on solid bedrock and therefore is reasonably stable. In contrast, the ice sheet in West Antarctica rests below the sea on bedrock and glacial till and is surrounded by floating ice pinned in by small islands buried below the ice. The ice on the mainland is unimaginably heavy and depresses the continental bedrock nearly 2,000 feet. The ice cap is so thick the region is practically devoid of earthquakes because the great weight of the ice prevents slippage along faults.

The largest ice sheet in the Northern Hemisphere lies on Greenland, the world's biggest island. Greenland separated from Eurasia and North America about 60 million years ago. About 8 million years ago, Greenland acquired a permanent ice cap in some places 2 miles or more thick. Snow precipitating from storm systems traversing across the surface of the glacier nourishes the Greenland ice sheet. The loss of ice at the boundary regions next to the ocean balances the glacier's growth. Large icebergs calving off glaciers entering the sea become a shipping hazard in the North Atlantic.

Some of the world's oldest rocks exist on Greenland. The Isua Formation in a remote mountainous area in the southwestern region comprises metamorphosed marine sediments formed about 3.8 billion years ago. Like Antarctica, Greenland lacks significant earthquake activity, supposedly because the weight of its massive ice sheet pressing down on the crust stabilizes existing faults, thus inhibiting fault slip.

Iceland is one of the coldest inhabited places on Earth and has the second largest ice cap in the Northern Hemisphere. Iceland is a broad volcanic plateau of the Mid-Atlantic Ridge, which rose above the sea some 16 million years ago. The abnormally elevated topography extends some 900 miles along the ridge, 350 miles of which lies above sea level. South of Iceland, the broad plateau tapers off, forming a typical midocean ridge. A mantle plume rising from the bottom of the mantle and lying beneath the plateau apparently augments the normal volcanic flow of the Mid-Atlantic Ridge, making possible Iceland's existence.

What makes the island so unique is that it straddles a spreading ridge system, where the two plates of the Atlantic Basin and adjacent continents pull apart. A steep-sided, V-shaped valley runs northward across the entire length of the island and is one of the few expressions of a volcanic rift system on land, with many volcanoes flanking the rift. Volcanism on Iceland produces glacier-covered volcanic peaks up to a mile high and generates intense geothermal activity.

Although Iceland is fortunate to possess an abundant supply of energy for electrical power and heating, it is not without its dangerous side effects, and frequent volcanic eruptions plague the island. The most destructive eruption in modern times buried much of the fishing village of Vestmannaeyjar on the island of Heimaey in 1973 (Fig. 117). An underglacier eruption in the sparsely populated southeastern part of the country in 1996 created massive flooding, as gushing meltwaters and icebergs raced 20 miles to the coast, destroying a major highway system linking the ice-covered island.

GLACIAL EROSION

A large portion of the landscape in the northern latitudes was sculpted by massive ice sheets that swept down from the polar regions during the last ice age. The glaciation was so pervasive that glaciers 2 miles or more thick enveloped much of upper North America and Eurasia. In some places, the crust was scraped completely clean of sediments, exposing the raw basement rock below and erasing the entire geologic history of the region.

Figure 117 *Houses partially buried by tephra in the eastern part of Vestamannaeyjar from the July 1973 eruption of Heimaey, Iceland.*

(Courtesy USGS)

One of the most remarkable examples of the power of glacial erosion is the Coteau des Prairies, a 200-mile-long by 70-mile-wide delta-shaped landform in eastern South Dakota. It is a low-relief formation created from a deposit of hard quartzite (metamorphosed quartz sandstone) that split the southward-flowing glacier of the last ice age into two lobes. The ice scoured the lowlands on either side, but did not overtop the Coteau itself, leaving it standing alone above the adjacent terrain.

Glaciers are the most effective agents of erosion in mountainous regions of the world with vast expanses of exposed rock upon which to work (Fig. 118). Mountain glaciers such as those in the Alps and Himalayas aggressively attack mountains, becoming perhaps the most potent erosional agents. In addition, mountains evolve their own climate as they grow, increasing the snowfall, thereby sowing the seeds of their own destruction.

During the ice ages, continent-wide glaciers buried many mountains in the northlands of Eurasia and North America, where glacial ice linked the Rocky Mountains with ranges in northern Mexico. In the Southern Hemisphere, small ice sheets expanded in the mountain ranges of Australia,

New Zealand, and the Andes of South America. Throughout the world, alpine glaciers capped mountains that are presently ice free.

Massive glaciers excavated some of most monumental landforms, and outwash streams from glacial meltwater carved out many peculiar landscapes. The glaciers descended from the mountains and spread across most of the northern lands, running over the terrain like an icy bulldozer. The power of glacial erosion is well demonstrated by the presence of deep-sided valleys carved out of mountain slopes (Fig. 119) by flowing ice a mile or more thick. Glacial erosion radically modified the shapes of stream valleys occupied by glaciers. The process is most active near the head of a glacier, where the ice deepens and flattens the gradient of the valley.

Glacial erosion reduces the land surface by glacier-transported rock fragments, bedrock scouring, and the erosive action of meltwater streams. Most glacial erosion involves the removal of rock by plucking, with abrasion smoothing and polishing the resulting form. Small hills or knobs in valleys overridden by glaciers are rounded and smoothed by abrasion.

Glaciers flowing down mountain peaks gouged large pits called cirques (Fig. 120), from French meaning "circle." They are semicircular basins or indentations with steep walls high on a mountain slope at the head of a val-

Figure 118 *Chickamin Glacier on the eastern slopes of Dome Peak, Glacier Peak Wilderness, Skagit County, Washington.*

(Photo by A. Post, courtesy USGS)

Figure 119 *A glacial valley, Dolores County, Colorado.*

(Photo by W. Cross, courtesy USGS)

ley. Cirque walls are cut back by the disintegration of rocks lower down the mountainside. The rock material embedded in the glacier gouges a concave floor. The expansion of adjacent cirques by glacial erosion creates aretes, horns, and cols. An arête, from French meaning "fish bone," is a sharp crested, serrated, or knife-edged ridge that separates the heads of abutting cirques. It also forms a dividing ridge between two parallel valley glaciers. A col is a sharp-edged or saddle-shaped pass in a mountain range formed by the headward erosion where cirques meet or intercept each other. When three or more cirques erode toward a common point, they form a triangular peak called a horn (Fig. 121).

As a glacier moves over bedrock, the ice abrades it by glacial scouring from the action of grinding or rasping. The abrasive agent is rock material dragged along by the glacial ice. Because of its fluid nature, ice does not erode the rock itself. Instead, it plucks fragments of bedrock by the plastic flow of the ice around them, with the rocks becoming part of the moving glacier. Boulders embedded in the ice gouge deep cuts into the easily eroded bedrock. Smaller rocks cut parallel striations or scratches in the bedrock, and finer material polishes it to a smooth finish.

When the last ice age ended, massive floods raged across the land, as water gushed from huge reservoirs trapped beneath the melting glaciers. Water flowing under the ice surged in vast turbulent sheets that scoured deep grooves in the crust, forming steep ridges out of solid bedrock. Torrents of meltwater laden with sediment surged along the Mississippi River to the Gulf of Mexico, significantly widening its channel and depositing new soil. Many other rivers overreached their banks to carve out broad floodplains.

Figure 120 *A hanging cirque west of Siyeh Glacier, Glacier National Park, Glacier County, Montana.*

(Photo by H. E. Malde, courtesy USGS)

Figure 121 *Cirques in the vicinity of Mount MacDonald, Mission Range, showing horn at the peak, Lake County, Montana.*

(Photo by C. D. Walcott, courtesy USGS)

GLACIAL DEPOSITS

Geologic evidence suggests that at least four major periods of glaciation occurred in Earth's history (Table 13). Most of the evidence for extensive glaciation is found in moraines and tillites, which are deposits of glacial rocks. Moraines are accumulations of rock material carried by a glacier and deposited in a regular, usually linear pattern that makes a recognizable landform. The sediment ranges in size from sand to boulders and shows no sorting or bedding normally associated with flowing water.

Moraines are named according to their position in relation to the glacier. A ground moraine is an irregular carpet of till, mainly composed of clay, silt, and sand deposited under a glacier. It is the most prevalent type of continental glacial deposit. A terminal moraine (Fig. 122) is a ridge of erosional debris deposited by the melting forward margin of a glacier that paused long enough for till to accumulate. It is a ridgelike mass of glacial debris formed by the foremost glacial snout and deposited at the outermost edge of glacial advance.

Tillites are a mixture of boulders and pebbles in a clay matrix consolidated into solid rock. They were deposited by glacial ice and are known to exist on every continent. In the Lake Superior region of North America, tillites are 600 feet thick in places and range east to west for 1,000 miles. In

TABLE 13 THE MAJOR ICE AGES

Time (Years)	Event
10,000–present	Present interglacial
15,000–10,000	Melting of ice sheets
20,000–18,000	Last glacial maximum
100,000	Most recent glacial episode
1 million	First major interglacial
3 million	First glacial episode in Northern Hemisphere
4 million	Ice covers Greenland and the Arctic Ocean
15 million	Second major glacial episode in Antarctica
30 million	First major glacial episode in Antarctica
65 million	Climate deteriorates; poles become much colder
250–65 million	Interval of warm and relatively uniform climate
250 million	The great Permian ice age
700 million	The great Precambrian ice age
2.4 billion	First major ice age

Figure 122 *Terminal moraine at the margin of a glacier, Deschutes County, Oregon.*

(Photo by I. C. Russell, courtesy USGS)

northern Utah, tillites 12,000 feet thick provide evidence for a series of ice ages following one after another in quick succession.

Similar tillites were found among Precambrian rocks in Norway, Greenland, China, India, southwest Africa, and Australia. In Australia, Permian marine sediments were found interbedded with glacial deposits, and tillites were separated by seams of coal, indicating that periods of glaciation were interspersed with warm interglacial spells. In South Africa, the Karroo Series, consisting of a sequence of late Paleozoic lava flows, tillites, and coal beds, reaches a total thickness of 20,000 feet.

Much of the upper midwestern and northeastern parts of the United States were overrun by thick glaciers during the last ice age. Many areas were eroded down to the granite bedrock, and the debris was left in great heaps. The glacially derived sediments covered much of the landscape, burying older rocks under thick layers of till. Glacial till is nonstratified or mixed material comprising clay and boulders deposited directly by glacial ice. The boulders are generally angular with sharp edges because they experienced little or no river transportation, which abrades and rounds rocks. Basil till at the base of a glacier was usually laid down under it. Ablation till on or near the surface of a glacier was deposited when the ice melted. The surface cover of sediment possibly protected the glacier from the heat of the sun. Some sun-heated rocks on the surface might have sunk into the glacier, forming deep depressions in the ice.

The most recent glacial period is the best studied of all because the evidence of previous glaciations was erased by the last ice age, as ice sheets eradicated much of the landscape. In many areas, the ice stripped off entire layers of sediment, leaving behind bare bedrock. In other areas, older deposits were buried under thick deposits of glacial till, forming elongated hillocks aligned in the same direction called drumlins (Fig. 123). Drumlins are tall and narrow at the upstream end of the glacier and slope to a low, broad tail. The hills appear in concentrated fields in North America, Scandinavia, Britain, and other areas once covered by ice. Drumlin fields might contain as many as 10,000 knolls, looking like rows upon rows of eggs lying on their sides.

Drumlins are the least understood of all glacial landforms. They appear to have originated when ice sheets contorted or deformed wet sediments lying beneath their bases. The sediments in the interior of a drumlin often form complex, swirling layers, indicating moving ice stretched and sheared them. How they attained their characteristic oval shape remains a mystery. Perhaps the extensive drumlin fields of North America were created during cataclysmic flood processes during melting of the vast ice sheets.

A roche moutonnée, from the French meaning "fleecy rock," is similar to a drumlin. The term was applied to glaciated outcrops because they resemble the backs of sheep, thus prompting the name sheepback rock. Roche moutonnée is a glaciated bedrock surface with asymmetrical mounds of varying

Figure 123 *A drumlin south of Newark, Wayne County, New York.*

(Photo by G. K. Gilbert, courtesy USGS)

shapes. The up-glacier side has been glacially scoured and smoothly abraded. The down-glacier side has steeper, jagged slopes, resulting from glacial plucking, an erosional process by which glaciers dislodge and transport fragments of bedrock. The fragments might have been pried loose by the plastic flow of the ice around them and became part of the moving glacier. The ridges dividing the two sides of a roche moutonnée are perpendicular to the general flow of the ice sheets. Such landforms are characteristic of glaciated Precambrian shields in the interiors of the continents.

Rugged periglacial regions existed at the margins of the ice sheets. Periglacial processes sculpted features along the tip of the ice and were directly controlled by the glacier. Cold winds blowing off the ice sheets affected the climate of the glacial margins and helped create periglacial conditions. The zone was dominated by such processes as frost heaving, frost splitting, and sorting, which created immense boulder fields out of what was once solid bedrock (Fig. 124).

Erratic boulders are glacially transported rocks embedded in glacial till or exposed on the surface. They range in size from pebbles to massive boulders and have traveled as far as 500 miles or more. The boulders were originally called drift because they appeared to have "drifted" in by flowing water or on floating ice. Today, the term *glacial drift* refers to all rock material deposited by glaciers or glacier-fed streams and lakes, where the greatest thicknesses are attained in buried valleys.

Figure 124 *Antonelli Glacier showing rugged periglacial area, including recessional and other moraines.*

(Photo by R. B. Colton, courtesy USGS)

Drift is divided into two types of material. One is till deposited directly by glacial ice and shows little or no sorting or stratification (layering) as though dumped haphazardly. The other is stratified drift, which is a well-sorted, layered material transported and deposited by glacial meltwater. Streams flowing from melting glaciers rework part of the glacial material, some of which is carried into standing bodies of water, forming banded deposits called glacial varves.

Erratics composed of distinctive rock types can be traced to their places of origin and indicate the direction of glacial flow. Indicator boulders are erratics of known origin used to locate the source area and the distance traveled for any given glacial till. Their identifying features include a distinctive appearance, unique mineral assemblage, or characteristic fossil content. Examples are erratics scattered from Iowa to Ohio, containing native copper torn from an outcrop in northern Michigan.

The erratics are often arranged in a boulder train, forming a line or series of rocks originating from the same bedrock source, and extend in the direction of glacial movement. A boulder fan is a conically shaped deposit containing distinctive erratics derived from an outcrop at the apex of the fan. The angle at which the margins diverge measures the maximum change in the direction of glacial motion.

Long, sinuous sand deposits called eskers (Fig. 125) formed out of glacial debris from outwash streams. They are winding, steep-walled ridges that can extend up to 500 miles in length but seldom exceed more than 1,000 to 2,000 feet wide and 150 feet high. Eskers were probably created by streams running through tunnels beneath the ice sheet. When the ice melted, the old stream deposits were left standing as a ridge. Eskers appear to have been deposited in channels beneath or within slow-moving or stagnant glacial ice. Their general orientation runs at right angles to the glacial edge. At the margin of a glacial lake, they might form river deltas. Some eskers originated on the ice and contained ice cores. Well-known esker areas are found in Maine, Canada, Sweden, and Ireland.

Like eskers, kames are found in areas where large quantities of coarse material are available during the slow melting of glacial ice. They are mounds composed chiefly of stratified sand and gravel formed at or near the snout of an ice sheet or deposited at the margin of a melting glacier. Meltwater must be present in sufficient quantities to redistribute the debris and deposit the sediments at the margins of the decaying ice mass. Most kames are low, irregularly conical mounds of roughly layered glacial sand and gravel that often occur in clusters. They accompany the terminal moraine region of both valley and continental glaciers and appear to represent sediment fillings of openings in stagnant ice. Many kames probably formed when streams flowed off the tops of glaciers onto the bare ground, dropping sediment into piles.

Figure 125 *An esker in Dodge County, Wisconsin.*

(Photo by W. C. Alden, courtesy USGS)

Glacial varves in ancient lake bed deposits are alternating layers of silt and sand laid down annually in a lake below the outlet of a glacier. Each summer when the glacial ice melted, meltwater turbid with sediments discharged into the lake, where the sediments settled out differentially, forming a banded deposit. The varying widths of ancient varves were thought to represent stages in the solar cycle when an increase in sunspot activity warmed the climate slightly, melting more ice than usual.

GLACIAL VALLEYS

The power of glacial erosion is well demonstrated by deep-sided valleys carved out of mountain slopes by thick sheets of flowing ice. A glacial valley is a river valley glaciated during the ice age. Many valleys were buried by ice a mile or more thick during the ice ages. Glaciers did not cut the original valley but only altered it by converting the formerly V-shaped valley into one that is U-shaped (Fig. 126), with a broad, flat bottom as much as 1,000 feet or more deep. A glacier straightens the valley it erodes because ice cannot turn as sharply as a river does due to its higher viscosity.

Glacial erosion removes ridges on the insides of curves within the stream valley and eliminates projecting spurs, which are ridges extending laterally

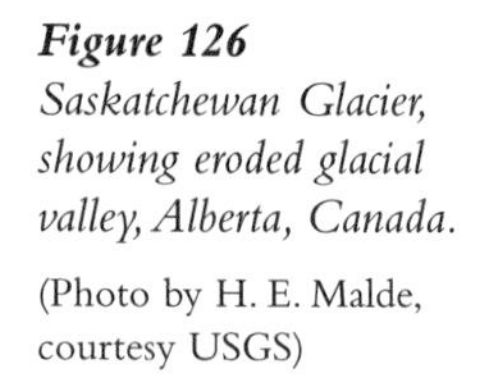

Figure 126
Saskatchewan Glacier, showing eroded glacial valley, Alberta, Canada.
(Photo by H. E. Malde, courtesy USGS)

Figure 127 *The Bridal Veil Falls, Yosemite National Park, Mariposa County, California.*

(Photo by F. E. Matthes, courtesy USGS)

from a mountain range. The glaciated valley floor is often irregular because ice more readily erodes fractured or weak rock, forming giant steps at intervals along the length of the valley. The principal river valley, whose source is near the mountain crest, contains a greater volume of ice than does a tributary stream valley. Consequently, the glacial ice erodes the main valley deeper than it does the tributary valley. After the ice melts, the tributary stream flows through a hanging valley above the main stream, into which it pours from a waterfall (Fig. 127).

Figure 128 *A glaciated ledge showing intensive striae marks the crest of the slope of a pre-glacial valley of the Nashua River, Worcester County, Massachusetts.*

(Photo by W. C. Alden, courtesy USGS)

As the glacial ice extended far down the valleys, it ground rocks on the valley floors. Rivers of solid ice with embedded rocks moved along the valley, grinding down the bedrock like a giant file as glaciers advanced and receded. The overriding ice also inscribed parallel furrows or striations on the valley floors as they sliced down the mountainsides. Glacial striae (Fig. 128) are fine-cut nearly parallel groves or scratches cut into the bedrock surface by rock fragments carved out by glaciers and cover large areas of northern Eurasia and North America. The striae are also engraved on the transported rocks and are excellent indicators of the direction of glacial flow. Glacial striae are usually found on Pleistocene deposits, but they also appear on rocks from earlier glaciations as far back as the Precambrian.

The glaciers also cut fjords, which are long, narrow, steep-sided inlets in glaciated mountainous coasts. During the last ice age, glaciers gouged deep fjords out of the coastal mountains of Norway, Greenland, Alaska, British Columbia, Patagonia in southern South America, and Antarctica. As a tidewater glacier on the coast erodes its valley floor below sea level, it cuts a steep-walled troughlike arm of the ocean. When sea levels returned to normal at the end of the ice age, the ocean invaded deeply excavated glacial troughs in the coastline. The side walls along the fjord are characterized by hanging valleys and tall waterfalls.

GLACIAL LAKES

Around 13,000 years ago, a gigantic ice dam lying on the border between Idaho and Montana held back a huge lake hundreds of miles wide and up to

2,000 feet deep. When it suddenly burst, the waters gushed toward the Pacific Ocean, and along their way they carved out one of the strangest landscapes the planet has to offer, known as the Channeled Scablands (Fig. 129). In the Altai Mountains of southern Siberia, perhaps the greatest flood ever to wash over the Earth was unleashed during the melting of the great ice sheets around 14,000 years ago, near the end of the last ice age. A glacier cutting across the Chuja Valley created a thick ice dam, which held back a large lake nearly 3,000 feet deep that contained some 200 cubic miles of water. When the ice dam broke and the lake burst through, a tremendous deluge of glacial meltwater rushed into the narrow river valley.

Lake Agassiz formed in a bedrock depression at the edge of the retreating ice sheet in south Manitoba, Canada. It was a vast reservoir of meltwater that was much larger than any of the existing Great Lakes. Similar large reservoirs of meltwater included Lake Lahontan in western Nevada, which expanded 10 times wider than today's remnant, and Lake Bonneville, which covered more than 20,000 square miles of Utah and Nevada and is now a dry salt flat punctuated by the Great Salt Lake, the only remaining body of water. The salt pan of Death Valley is the evaporated remains of a series of lakes, the largest of which was Lake Manly, whose basin was filled with runoff from the glaciated Sierra Nevada between 75,000 and 10,000 years ago.

Figure 129 *Palouse Island near the eastern margin of the Channeled Scablands, Washington.*

(Photo by F. O. Jones, courtesy USGS)

When the North American ice sheet retreated at the end of the last ice age, most of its meltwater flowed down the Mississippi River. Sometimes, huge lakes of meltwater trapped beneath the ice sheet broke free and rushed down river valleys in torrents to the Gulf of Mexico and the Atlantic Ocean equal to the flow of several Mississippi Rivers. While flowing under the ice, water surged in vast turbulent sheets scouring deep grooves in the surface, forming steep ridges carved out of bedrock. Several times, massive surges of meltwater broke loose to further gouge the landscape.

When the ice sheet retreated beyond the Great Lakes, which were themselves carved out by the glaciers, the meltwater took a separate route down the St. Lawrence River. The cold waters entered the North Atlantic, initiating a return to ice age conditions and a pause in the melting known as the Younger Dryas, named after an Arctic wildflower. Also during this time, the Niagara River Falls began cutting its gorge. It has traversed more than 5 miles northward since the ice sheet began to retreat, cutting into the bedrock at a rate of up to 3 feet per year.

Dotting much of the northern lands are glacial lakes developed from major pits excavated by the glaciers. Smaller lakes were formed when a large block of ice buried by glacial outwash sediments melted, leaving depressions called kettles (Fig. 130). They originated from large blocks of ice buried by glacial outwash sediments. The depressions are circular or elliptical because ice blocks round out as they melt. The depressions range up to 10 miles or more in diameter and up to 100 feet or more deep. Not all kettles contain water, however, and some dry kettles hold a stand of trees that gradually fall off below ground level.

Figure 130 *A kettle hole in gravels near the terminus of Baird Glacier, Thomas Bay, Petersburg district, southeastern Alaska.*

(Photo by A. F. Buddington, courtesy USGS)

Most kettles occur singly or in groups. A collection of large numbers of kettles produces a terrain in the shape of basins and mounds, called knob and kettle topography. The landscape features an irregular assemblage of knolls, mounds, or ridges between depressions or kettles and sometimes contain swamps or ponds. The undulating landform is a type of terminal moraine, possibly created by slight oscillations of an ice front as it recedes. A section of knob and kettle topography called hummocky moraine develops either along a live ice front or around masses of stagnant ice.

The largest of the glacial lakes are the Great Lakes, bordering between the United States and Canada. The lakes presently receive huge quantities of sediments derived from the continent, and the constant buildup gradually makes them shallower. In some future date, the lakes will completely fill with sediment and became dry, flat, featureless plains, until such time when the ice sheets return to scour out their basins once more.

SURGE GLACIERS

Some 200 surge glaciers in North America are heading toward the sea. During most of its life, a surge glacier behaves normally, moving along at a snail's pace of perhaps a couple of inches a day. At regular intervals of 10 to 100 years, however, the glaciers gallop forward upward of 100 times faster than their normal speed. One dramatic example is the Bruarjokull Glacier in Iceland, and in a single year it advanced 5 miles, at times achieving speeds of 16 feet an hour. For 85 years, Hubbard Glacier (Fig. 131) had been flowing toward the Gulf of Alaska at a steady rate of about 200 feet per year. However, in June 1986, the 80-mile-long river of ice surged ahead as much as 46 feet in a single day. In recent years, it has been retreating, possibly due to a spell of global warming.

A surge often progresses along a glacier like a great wave, proceeding from one section to another toward the sea. The reason for glacial surge has not yet been fully explained, although they might be influenced by the climate, volcanic heat, and earthquakes. However, surge glaciers exist in the same regions as normal glaciers, often almost side by side. For example, the great 1964 Alaskan earthquake failed to produce more surges than normal.

In Antarctica, large flat areas beneath glaciers are thought to be subglacial lakes, kept from freezing by the interior heat of the Earth. The temperature a mile below the surface of the ice can be warmer than the temperature of the ice on top. Add to that the high pressures that occur at such depths, and liquid water can exist several degrees colder than its normal freezing point. The pools of liquid water tend to lubricate the ice streams, helping them flow down the mountain valleys to the sea, where they calve off to form icebergs.

Figure 131 *The Hubbard Glacier, Russell Fiord, Yakutat district, Alaska Gulf region, Alaska.*

(Photo by Austin Post, courtesy USGS)

This enables ice streams up to several miles broad to glide smoothly along the valley floors.

Behind a wall of mountains that form the spine of Antarctica, the Transantarctic Range, rivers of ice slowly flow outward and down to the sea on all sides. The ice escapes through mountain valleys to the ice-submerged archipelago of West Antarctica, and to the great floating ice shelves of the Ross and Weddell Seas. West Antarctica is traversed by ice streams several miles broad, and rivers of solid ice flow down mountain valleys to the sea. Muddy pools of meltwater lie on the bottom and lubricate the ice streams, allowing them to glide smoothly along the valley floors.

The banks and interior portions of the ice streams are marked by deep crevasses. A glacial crevasse is a crack or fissure in a glacier, resulting from stress due to movement. They are generally several tens of feet wide, a hundred or more feet deep, and up to 1,000 or more feet long. The banks of glaciers are often flanked by deep crevasses, where they meet the walls of the glacial valley. Crevasses also run parallel to each other down the entire length of the ice streams, especially when the central portion of the glacier flows faster than the outer edges. Snow bridges occasionally span the crevasses, occasionally hiding them from view. Sometimes a stream of meltwater can be heard gurgling far below from open crevasses slicing through a glacier.

PATTERNED GROUND

Among the most barren environments on Earth is the Arctic tundra of North America and Eurasia. The tundra covers about 14 percent of the Earth's land surface in an irregular band that winds around the top of the world, north of the tree line and south of the permanent ice sheets. Most of the ground in the Arctic tundra is permafrost and frozen year round, and only the top few inches of the soil thaws during the short summers.

Although the ground is bathed in 24-hour sunlight during the short summer, the soil temperature seldom reaches above freezing because as ice thaws it absorbs heat from its environment. The Arctic tundra is also one of the most fragile regions, and even small disturbances such as vehicle tracks from oil exploration activities can cause a great deal of damage, taking decades to repair.

In many parts of the tundra, soil and rocks are fashioned into strikingly beautiful and orderly patterns that have confronted geologists for centuries. Every summer, the retreating snows in the Arctic tundra unveil a bizarre assortment of rocks arranged in a honeycomblike network as the ground begins to thaw, giving the landscape the appearance of a tiled floor (Fig. 132). These patterns are found in most of the northern lands and alpine regions, where the soil is exposed to moisture and seasonal freezing and thawing. The polygons range in size from a few inches across when composed of small pebbles to 30 to 50 feet wide when large boulders form protective rings around mounds of soil.

Figure 132 *Patterned ground on the northern Alaskan sea coast near Barrow, Barrow district, Alaska.*

(Photo by T. L. Pewe, courtesy USGS)

The regular polygonal patterned ground in the Arctic regions might have been formed by the movement of soil of mixed composition upward toward the center of the mound and downward under the boulders, making the soil move in convective cells. The coarser material composed of gravel and boulders is gradually shoved radially outward from the central area, leaving the finer materials behind. The arrangement of rocks in this manner suggests that the soil is being churned up by convection.

Other geometric designs in the Arctic soil include steps, stripes, and nets, which lie between the circles and polygons. These other forms can reach 150 feet in diameter. Relics of ancient surface patterns measuring up to 500 feet have been found in former permafrost regions. Even on Mars, spacecraft have revealed images of furrowed rings, polygonal fractures, and ground-ice patterns of every description. These among many other fascinating features presented in the Arctic make the region one of the most interesting places on the face of the Earth.

After a discussion of glacial features, the next chapter takes a journey underground and examines caves and related structures.

9

CAVES AND CAVERNS

DELVING BENEATH THE EARTH'S SURFACE

This chapter examines underground structures including caves and related features. Possibly no other geologic structure captures the popular imagination more than caves. They are, after all, thought to be our ancestral homes. Early geologists believed that the continents were undermined by huge subterranean caverns that would collapse and flood the land with seawater, resulting in the Great Flood. Caves are also misunderstood, and a great deal of superstition surrounds them.

Some caves are extensive mazes where one could get lost. Many caves are homes to countless numbers of bats, which only exacerbates the misconceptions. People are simply afraid of caves or become claustrophobic due to the dark, enclosed spaces. Regardless of these fears, spelunkers, or cavers as they call themselves, have an insatiable zeal for caves and are impatient to explore farther into the bowels of the Earth.

CAVE FORMATION

Caves are the most spectacular examples of the handiwork of groundwater. The dissolving power of water is well demonstrated by the formation of caves in soluble carbonate rock such as limestone. Although limestone caves are the most common, caves also can form in dolomite and gypsum. Calcite is a calcium carbonate and the most common mineral in limestone. It is soluble in water that contains dissolved carbon dioxide. Another form of calcite, called travertine, is a common mineral found in caves and hot carbonate springs (Fig. 133).

Rainwater filtering through sediment layers reacts with carbon dioxide to form a weak carbonic acid. In addition, if the overlying formation contains pyrite, the sulfur in the mineral could be oxidized by rainwater and converted into sulfuric acid. The acid is carried downward through cracks in the underlying rock layers and dissolves calcite or dolomite, the minerals that constitute limestone and dolostone. This action enlarges the fissures and eventually creates a path for more acidic water. Most of the dissolving occurs near the top of the water table, where a mixture of groundwater and water percolating from above dissolves rock more readily than either form of water alone.

The wide variety in the shape and structure, or morphology, of cave passages results from the chemical reactions that dissolve limestone and the position of the base level. The base level is the level at which water flows in an area and is influenced by the elevation of a nearby stream or river. Cave passages constructed near the top of the water table are affected by changes in base level. It also plays an important role in their orientation, even in areas where the landscape is more complex.

Figure 133 *Mammoth Hot Springs, Yellowstone National Park, Wyoming.* (Photo by W. H. Jackson, courtesy USGS)

Different levels of passages in a cave develop at different times, when groundwater changes base levels. A lower level might begin to form below the water table, while an upper level forms near the top of the water table. Vertical shafts, up to several hundred feet deep, called domepits, connect the various levels. They develop relatively late in the cave-forming process, as rainwater trickles down toward the lower water table.

Cave passages can be affected by different geologic conditions, such as the existence of folds and faults at different parts of the cave. The greatest variety of cave morphology occurs where rock layers are folded. In some areas, rock layers were folded and pushed downward to form synclines. Other rock layers were folded and bulged upward to form anticlines. Many caves formed on the sides of synclines and anticlines have straight, parallel passages that run along the axis of the fold in the direction of the formation strike, which runs right angle to the direction of the structural dip, or downslope. Most caves form on only one level, but many near the tops of mountains have several parallel levels of passages. These passages are connected by shorter ones that run along the slopes of the folded sedimentary beds.

The stratigraphy of horizontal rock layers is another important factor to cave development. The passage orientation and morphology are largely controlled by geologic, hydrologic, and structural conditions. When a cave forms near the axis of a fold, a maze of passages might develop. The axes of synclines and anticlines are usually more fractured than the sides of the folds and transmit limestone-dissolving acidic water more easily. Joints, or fractures in the rock, are also important to the development of cave passages in areas where rock layers are not folded or tilted.

Caves also develop in the zone of seasonal water table fluctuation by dissolution of limestone along joint planes. They form from underground channels that carry out water that seeps in from the water table. This creates an underground stream similar to how streams form on the surface by a breached water table. Caves also develop in sea cliffs (Fig. 134) by the ceaseless pounding of the surf or by groundwater flow through an undersea limestone formation hollowed out as the water empties into the ocean. Sea arches, such as Needle's Eye on Gibraltar Island in western Lake Erie, were created by wave action on limestone promontories with zones of differential hardness.

Some caves were formed by tectonic movements of large rock masses that make up the Earth's crust. Other caves are made by weathering processes such as rock falls. Smaller caves are fashioned out of sandstone. Several caves are bored into the red sandstone formations around Moab, Utah. Houses built in natural and artificial caves are well insulated and maintain fairly uniform temperatures year round.

Figure 134 *Sea cave cut into siltstone, Chinitna district, Cook Inlet region, Alaska.*

(Photo by A. Grantz, courtesy USGS)

KARST TERRAIN

Limestone and other soluble materials underlie large portions of the world. When groundwater percolates downward through these formations, it dissolves minerals, forming cavities or caverns. When the land overlying these caverns suddenly collapses, it forms a deep sinkhole (Fig. 135). The sinkholes are generally up to 300 feet or more wide and 100 feet or more deep. At other times, the land surface settles slowly and irregularly.

The pits created by dissolving soluble subterranean materials produce a pocked-marked landscape known as karst terrain. The name is derived from the region of Karst on the coast of Slovenia famous for its numerous caves. Karst terrain is generally found in areas with moderate to abundant rainfall. Throughout the world, some 15 percent of the land surface rests on karst terrain occupied by millions of sinkholes. Although the formation of sinkholes is a natural phenomenon, the process is often accelerated by the withdrawal or disposal of water.

The major locations for karst terrain and caverns in the United States are mainly in the southeastern and midwestern parts of the country. Karst terrain also covers some portions of the Northeast and West. In Alabama, where limestone and other soluble sediments overlie nearly half the state, thousands of sinkholes pose serious problems for highways and construction projects. About a third of Florida is underlain by eroded limestone at shallow depths, where sinkholes are a common sight.

The settlement of the land surface can cause extensive damage to structures built over cavities formed by dissolving soluble minerals. A dramatic example of this phenomenon occurred in Bartow, Florida, on May 22, 1967 when a sinkhole 520 feet long and 125 feet wide collapsed under a house (Fig. 136). On December 12, 1995 heavy rainfall and a sewer pipe break in San Francisco, California, created a huge sinkhole as deep as a 10-story building that swallowed a million-dollar house and threatened dozens more.

Karst plains are flat areas with karst features in regions of nearly horizontal limestone strata. A blind valley is a river valley in karst terrain that ends abruptly where the stream disappears underground, called a swallow hole. During heavy rains, a blind valley might become a temporary lake. A type of blind valley called a karst valley forms by the coalescence of several sinkholes. Often sinkholes fill with water and become small permanent lakes.

The shallow seas around the Bahamas, southeast of Florida, contain water-filled sinkholes that produce large dark pools of deep seawater called blue holes. They formed during the ice age when the ocean dropped several hundred feet, exposing sections of the ocean floor well above sea level. Acidic rainwater seeping into the ground dissolved the limestone bedrock and created vast subterranean caverns. Under the weight of the surface rocks, the roofs of the caverns collapsed, exposing huge gaping pits. When the glaciers melted at the end of the ice age, the area refilled with seawater as the ocean rose near its current level. Much fear and superstition surrounds blue holes because they

Figure 135 *A sinkhole in Minnekahta limestone, Weston County, Wyoming.*

(Photo by N. H. Darton, courtesy USGS)

Figure 136 *A sinkhole 520 feet long, 125 feet wide, and 60 feet deep that collapsed under a house in Bartow, Florida, on May 22, 1967.*
(Courtesy of USGS)

often exhibit strong eddy currents or whirlpools that can swamp small boats during incoming and outgoing tides.

The karst terrain in the jungles of Mexico's Yucatán Peninsula gives birth to a bizarre realm of giant underwater caves and sinkholes linked by long twisting passages. The sinkholes formed when the upper surface of a limestone formation collapsed, exposing the watery world deep beneath the jungle floor. The sinkholes provide access to a vast subterranean world 100 feet below the surface of the Earth.

The underlying limestone is honeycombed with long tunnels many miles long and huge caverns that could easily hold several houses. Like surface caves, the Yucatan caverns contain a profusion of icicle-shaped formations of stalactites hanging from the ceiling and stalagmites clinging to the floor. The formations also include delicate, hollow stalactites called soda straws that took millions of years to create but are destroyed in mere moments by careless divers. Creatures blinded by generations of unused eyesight live in the darkest recesses of the caves. Such caves represent an almost entirely new ecosystem, filled with unusual life forms.

Strange, previously unknown creatures, including spiders, beetles, leeches, scorpions, and centipedes inhabit the deep dark passages of Movile Cave 60 feet below ground in southern Romania. The cave is a closed subterranean ecosystem sealed off from the surface and nourished by hydrogen sulfide rising from the Earth's interior. Bacteria at the bottom of the food chain metabolize hydrogen sulfide in a process called chemosynthesis.

The cave's bizarre occupants evolved over the past 5 million years and live with little oxygen and absolutely no light. As a result, they lack pigmentation and eyesight. The cave, which winds beneath 150 square miles of dry countryside, began when the Black Sea dropped precipitously some 5.5 million years ago. The cave developed in a limestone formation when the waters began rising again. It was sealed off from the outside world when clay impregnated the limestone, making it watertight, and when thick layers of wind-driven sediment were deposited on top during the ice ages.

Other cave creatures consist of bacterial colonies that are apparently unique to a cavern located in the Mexican state of Tabasco about 40 miles south of Villahermosa. A white slime coated the cave walls and ceiling, which dripped with water more acidic than battery acid. The slime contained bacteria that ingested sulfur as its energy source and excreted a strong sulfuric acid. The bacteria serve as the bottom of the food chain for an unusual sulfur-based ecosystem inside the mile-long cave.

The biomass of the subterranean microbes could equal that of all organisms living on the surface. They might reach as far as 2.5 miles below the continental crust and more than 4 miles into the oceanic crust. Any deeper, and the rock is presumably too hot for life. Their existence also suggests life might even have originated deep down in the bowels of the Earth, well protected from the harsh conditions on the surface, where any life attempting to evolve above ground would have met a sizzling death.

Deep in a cave, bacterial spores were discovered entombed long ago in fossilized tree sap, or amber. Some microbes might be as old as 135 million years, having lived during the height of the dinosaur age. Even older spores of bacteria were found in salt crystals dating to 250 million years ago. In times of crises, many bacteria and fungi can transform themselves into biologic time capsules until conditions of life improve. As spores, the microbes cease moving, eating, and reproducing. They become virtually indestructible, able to survive over lengthy periods without air or water and can withstand the harshest environments.

NATURAL BRIDGES

The same processes that form caves also create natural bridges (Fig. 137). Natural bridges and arches are among the most fascinating geologic features the Earth has to offer. Natural bridges are narrow, continuous archways of rock that often span a ravine or a valley. Rainbow Bridge near Lake Powell on the border between Utah and Arizona is the world's largest natural bridge.

Figure 137 *Natural bridge in the Wasatch Formation at the south end of Bryce Canyon National Park, Utah.* (Photo by R. G. Luedke, courtesy USGS)

Over the years, sections have broken off and fallen to the ground, prompting fears that in its weakened state it might collapse altogether. It is an important Native American ceremonial site, and visitors are asked not to walk under the great arch.

Natural bridges are of complex origin and form by a variety of processes. They are the product of erosion and weathering of resistant rocks, such as sandstone or limestone, that contain layers that resist erosion in varying degrees. Some layers are hard and resist chemical and mechanical weathering, while others are easily weathered and eroded. If a resistant layer lies above a softer layer of rock, it forms a protective cap. When a vertical joint or fracture penetrates the softer rock and water flows through this structure, the softer rock erodes and undercuts the resistant capping layer.

Some natural bridges that formed in sandstone represent rock-shelter caves where part of the roof has collapsed, either by large blocks of rock breaking away or by a slower process of spalling, or chipping away grain by grain. When a narrow part of the roof is left intact, a bridge is formed. The roof collapse generally takes place in a section bounded by joints. Water from a surface stream might be captured in joints that penetrate the formation and flows through the rock. Eventually, the flowing water widens the joints and cuts below the uppermost layer of rock, resulting in a natural bridge.

Natural bridges formed in limestone or dolomite are formed by the chemical and mechanical weathering of an underlying, less-resistant layer of rock. Many of these bridges develop on narrow ridges. Generally, the opening beneath the bridge has been widened by solution weathering, with water percolating from the surface along a vertical joint. The passage beneath the bridge span is in effect a segment of a limestone cavern, formed by solution along a joint exposed at either end as the ridge is narrowed by erosion.

Another type of natural bridge is formed when a large, detached block of rock falls or tilts so that it bridges the gap between two other blocks of rock. In limestone terrain, natural bridges are created in tunnels excavated by groundwater solutions, resulting in a collapse of the tunnel roof. Natural Bridge in Virginia (Fig. 138) is the most famous example of this type of bridge in the United States. Many arches fashioned by the lateral erosion of a stream flowing around and eventually through the rock evolved into natural bridges. Other small bridges or arches are associated with sink holes. Even petrified tree trunks are known to construct natural bridges. Snow bridges often span crevasses in glaciers.

Across many arid lands are magnificent stone archways (Fig. 139). The term *arch* refers to a span of rock that has no natural stream flowing beneath it. Arches are especially plentiful in Arches National Monument near Moab, Utah, and are a delight to a large number of visitors to the park. The arches formed when rock eroded at different rates due to a variance in resistance to erosional forces. They were created partly by wind erosion of thick sandstone beds. Rainwater first loosens the sand near the surface, while wind removes the loose sand grains. Weathering by rainwater in combination with wind erosion then abrades the rock, cutting through the weakened formation similarly to sand blasting.

LIMESTONE CAVES

Limestone is formed by biologic and chemical precipitation of carbonaceous minerals dissolved in seawater. Carbonic acid, produced by the chemical reaction of water and carbon dioxide in the atmosphere, dissolves calcium and silica minerals from rocks on the surface to form bicarbonates. The bicarbonates enter rivers that empty into the ocean and are mixed with seawater and precipitate by biologic activity and direct chemical processes. Organisms use the calcium bicarbonate to build their shells and skeletons composed of calcium carbonate. When the organisms die, their skeletons fall to the bottom of the sea, where the calcium carbonate as a calcite ooze builds thick deposits of carbonate rock.

The most common carbonate rock is limestone, which constitutes about 10 percent of all surface rocks. It is generally deposited by biologic processes, as evidenced by abundant marine fossils in limestone beds. Some limestone is chemically precipitated directly from seawater, and minor quantities precipitate in evaporite deposits. Dolomite resembles limestone, but is produced by the replacement of calcium in limestone with magnesium. The mineral is more resistant to acid erosion than limestone, which accounts for the impressive Dolomite peaks of Europe.

Chalk is a soft, porous carbonate rock that erodes easily. Thick chalk beds were laid down during the Cretaceous, which is how the period got its name.

Figure 138 *Natural Bridge, Rockbridge County, Virgina.*

(Photo by J.K. Hillers, courtesy USGS)

Figure 139 *Gothic Arch in Navajo sandstone resting on Kayenta sandstone, Garfield county, Utah.*

(Photo by H. E. Gregory, courtesy USGS)

Creta is Latin for "chalk." The thick chalk banks on the Suffolk coast of England have been wearing away by wave action for centuries. A major storm at sea can erode the tall cliffs landward several tens of feet. Sometimes the pounding of the surf punches a hole in the chalk to form a sea arch.

Over a lengthy period, groundwater dissolves large quantities of limestone, forming a system of tunnels, large rooms, and galleries. The differences in their shapes are due to the geology, hydrology, and structure of the rocks where the caves form. Anvil Cave in Georgia comprises nearly 13 miles of passages in an area of only 18 acres. Many caves occupy more than one level, while some are entirely vertical. Ellison's Cave in Georgia has 10 miles of passages and more than 1,000 feet of vertical drop, making it one of the deepest caves. The Mammoth Cave system in Kentucky is the longest in the world, with more than 300 miles of known passages spread among six levels. Most of the passages on each level are along bedding planes that separate different rock layers.

Carlsbad Caverns in southeast New Mexico (Fig. 140) is formed in limestone that was once a large reef similar to Australia's Great Barrier Reef. The main portion of the cave is composed of massive limestone that remains much the same throughout the formation. The second part is layered and fashioned out of pieces of rock that eroded from the main part of the limestone and from overlying formations. The three largest caves formed at the area where the two formations meet, due to a layer of impermeable rock that lies between them. The many levels resulted from changes in groundwater base level during the cave's development.

Figure 140 *Hattin's Dome and Temple of the Sun at Carlsbad Caverns National Park, Eddy County, New Mexico, 1923.*

(Photo by W. T. Lee, courtesy USGS)

LAVA CAVES

Lava is molten rock, or magma, that reaches the throat of a volcano or fissure and flows onto the surface with little or no explosive activity. Lava flows are generally tabular igneous bodies that are thin compared with their horizontal extent. Much of the shape of a lava flow is determined by the terrain upon which it flows. On flat plains, lava flows are usually horizontal, whereas on the slopes of volcanoes they might consolidate into considerable thickness. Lava flows greater than 300 feet thick are rare. In Hawaii individual flows average 10 to 30 feet in thickness.

The magma from which lava is produced has a low viscosity, allowing volatiles and gases to escape with comparative ease. This produces much quieter and milder eruptions such as those of the Hawaiian volcanoes. Lava is largely composed of basalt that is about 50 percent silica, dark in color, and quite fluid. Pahoehoe, or ropy lavas, are highly fluid basalt flows produced when the surface of a flow congeals to form a thin plastic skin. As the melt beneath continues to flow, it molds and remolds the skin into billowing or ropy-looking surfaces. When the lavas eventually solidify, the skin retains the appearance of the flow pressures exerted on it from below.

Highly fluid lava moves rapidly, especially down the steep slopes of a volcano. The flow rate is also determined by the viscosity and the time the lava

Figure 141 *The Bridge of the Moon lava tube, Craters of the Moon National Monument, Idaho.*

(Photo by H. T. Stearns, courtesy USGS)

takes to harden. Most lavas flow at a walking pace to about 10 miles per hour. Some lava flows have been clocked at only a snail's pace, while others run as fast as 50 miles per hour. Some very thick lavas creep along slowly for months or even years before solidifying.

If a stream of lava hardens on the surface and the underlying magma continues to flow away, a long tunnel, called a lava tube or cave, is formed. It can reach several tens of feet across and extend for hundreds of feet. In exceptional cases they might extend for up to 12 miles in length. Excellent examples of lava caves can be found in the Modoc lavas in northeastern California and Craters of the Moon in Idaho (Fig. 141). The caves might be partially or completely filled with pyroclastic material or sediments that washed in through small fissures. Sometimes the walls and roof of the lava caves are adorned with stalactites, and the floor is covered with stalagmites composed of deposits of lava.

In volcanic terrain in parts of Alaska, California, Oregon, Washington, and Hawaii, volcanic-related subsidence is usually caused by local collapses above shallow tunnels. If the roof of a lava tunnel collapses, it leaves a circular or elliptical depression on the surface of the lava flow. A good example is a lava flow in New Mexico, where a collapse depression is nearly a mile long and 300 feet wide. Rilles on the moon are trenchlike valleys that resemble collapsed lava tubes (Fig. 142). Lava tubes on Earth are very similar to rilles on the lunar surface as well as on other planets and moons.

ICE CAVES

Iceland is a surface expression of the Mid-Atlantic Ridge and straddles both sides of the rift, where volcanic eruptions are quite frequent. Icelandic vol-

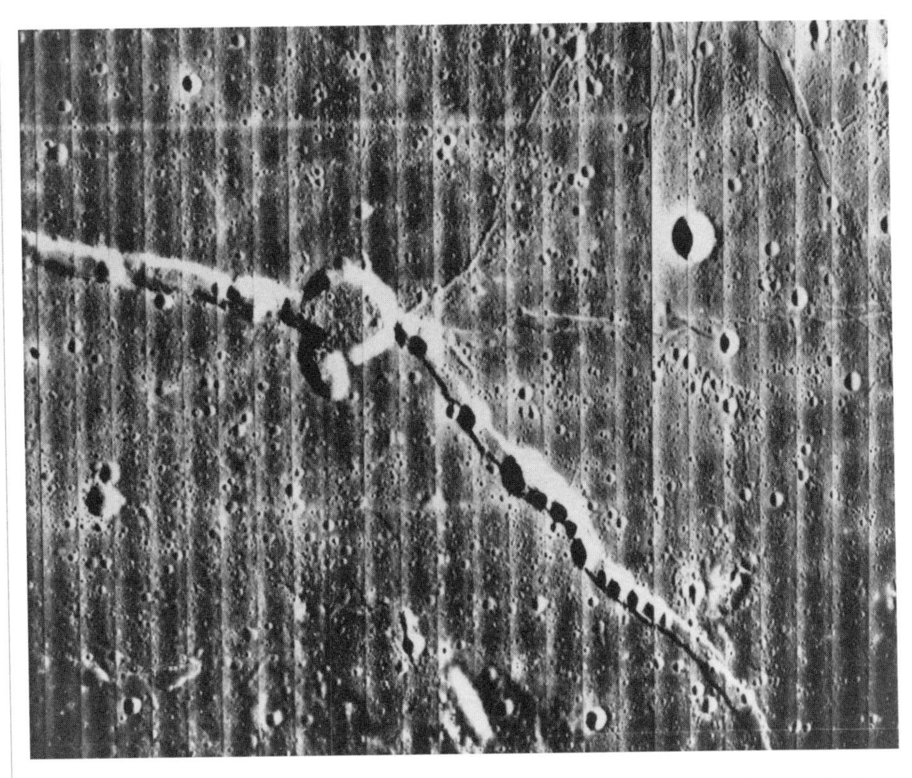

Figure 142 *The lunar crater Hyginus and Hyginus Rille.*

(Photo by D. H. Scott, courtesy USGS and NASA)

canism produces glacier-covered volcanic peaks up to 1 mile high. In 1918, an eruption under a glacier unleashed a flood of meltwater, called a glacier burst. In a matter of days, the underglacier eruption released up to 20 times more water than the flow of the Amazon, the world's largest river.

A similar under-ice eruption on September 30, 1996 melted through the 1,700-foot-thick ice cap and sent massive floodwaters and icebergs dashing to the sea a month later, for a short while forming the second largest river in the world. It destroyed three bridges, telephone lines, and the only highway running along Iceland's southern coast, with damages estimated at $15 million. At least 13 underglacier eruptions have occurred during the last half of the 20th century. These massive floods of meltwater, known as jokulhlaups, are well known to Icelanders since the 12th century.

A glacier burst is a sudden release of meltwater from a glacier or subglacial lake. Water accumulates in depressions within the ice margins and erupts through the ice barrier, sometimes resulting in a catastrophic flood. The process of accumulation and release might occur at almost regular intervals. The phenomenon is most common in Iceland, where it accompanies volcanic or fumarolic (volcanic steam) activity. Geothermal heat beneath the ice creates a large reservoir of meltwater as much as 1,000 feet deep. A ridge of rock acts as a dam to hold back the water. When the dam suddenly breaks open, the flow of water forms a channel under the ice that is upward of 30 miles long.

The gigantic ice sheets atop Antarctica hide many volcanoes, and under-ice eruptions as much as a mile or more beneath the glacier can produce massive floods of meltwater. Several volcanoes puncture the ice of West Antarctica. Many of Antarctica's dormant volcanoes are buried within the ice, and extensive volcanic deposits underlie the ice sheets. When active volcanoes erupt beneath the ice, they spout great floods of meltwater, which mixes with the underlying sediment, forming glacial till tens of feet thick.

A round depression in the ice of the Antarctic Ross Ice Shelf was some 4 miles wide and 160 feet deep. Only an active volcano erupting under the glacier could have melted such a large area of ice. Using radar to penetrate the ice sheet, a 4-mile-wide, 2,100-foot-high volcano was discovered under more than 1 mile of ice. The volcano sits in the middle of a giant caldera 14 miles wide within a rift valley, where the Earth's crust is being stretched apart and hot rock from the mantle is rising to the surface. Water gushing from such an under-glacier eruption produces a glacier burst. Satellite images have revealed other circular depressions in the ice, suggesting many more volcanoes are lurking beneath the glaciers.

Meltwater flowing out of a glacier often carves through the ice to form an ice cave (Fig. 143) that can be followed upstream for long distances. Often the stream can be heard running beneath the glacier from crevasses slicing

Figure 143 *An ice cave on the Ross Ice Shelf, Antarctica.*

(Photo by W. J. Collins, courtesy U.S. Navy)

through the ice. The swift-flowing stream carries a heavy load of sediment deposited at the mouth of the glacier (Fig. 144). This forms glacial varves, alternating layers of silt and sand laid down annually in a lake below the outlet of a glacier. Each summer when the glacial ice melts, turbid meltwater discharges into the lake and sediments settle out differentially, forming a banded deposit.

Long, sinuous sand deposits, called eskers, were formed out of glacial debris from outwash streams. They are winding, steep-walled ridges up to several hundred miles in length and up to 1,000 feet or more wide. Eskers are thought to have been created by streams running through tunnels beneath an ice sheet during the last ice age. When the ice melted, the old stream deposits remained standing as a ridge.

Since the Little Ice Age, between 1430 and 1850, when global temperatures were as much as 2 degrees Celsius cooler than they are today, outflow glaciers have been retreating. Their farthest advance is marked by terminal moraines composed of glacial deposits laid down at the leading edge of a glacier. Also, various limits of retreat can be identified by studying the growth of primitive plants called lichens on the rocks, whose research is known as the science of lichenometry.

CAVE DEPOSITS

A cave's elaborate architecture is determined mainly by water filtering in from above. Rain and melting snow on the surface seep into the cave through thick deposits of soil and rock. As the water percolates downward, it picks up carbon dioxide from decaying plants to form a weak solution of carbonic acid. The acidic water trickles downward, dissolving limestone on its way. Upon reaching the surface of a cave, the water is exposed to the air and releases carbon dioxide, which precipitates calcite. Another form of calcite, called travertine, is common in hot carbonate springs and in caves. Parallel-banded travertine makes an excellent decorative or architectural material. It is usually deposited from cold water solutions often found as stalactites and stalagmites in caves.

Limestone caves contain long "icicles" of calcite that grow by the precipitation of acidic groundwater seeping through the rock (Fig. 145). A drop of water hanging from the ceiling of a cave when exposed to the air loses some of its acidity and can no longer hold calcite in solution. When the drop falls to the floor, a tiny calcite crystal is left behind. Over time, more drops gather at the same spot, and bit by bit the calcite grows downward from the ceiling, forming a stalactite, also called dripstone because it formed by dripping water containing dissolved calcium carbonate.

Water drops falling from the stalactite onto the cave floor still contain a small amount of dissolved calcite. When the drop hits the floor of the cave, it

Figure 144 *Meltwater from glaciers in Glacier National Park, Montana.*
(Courtesy National Park Service)

splatters and loses more acidity, precipitating another tiny calcite crystal. As more drops deposit calcite, a stalagmite grows upward toward the overhanging stalactite. Sometimes the two formations meet to form columns. The process is extremely slow, and stalactites and stalagmites take hundreds of years to grow a single inch.

Figure 145 *Mirror Lake in Grand Caverns, Augusta County, Virginia.*

(Photo by W. T. Lee, courtesy USGS)

Some caves also contain exquisite twisting fingers of calcite or aragonite called helictites. They form similarly to stalactites, but water drips through them too slowly for drops to form, creating contorted, branching cave deposits. The moisture reaches the tip of the helictite and evaporates, so that crystals do not grow straight down but in curls and spirals. Aragonite, which is similar to calcite but has a different crystal structure, forms rough, needlelike helictites.

Other cave deposits include cave coral, created when water reaches the cave through a network of channels too small for drops to form. Instead, moisture is squeezed out onto the cave wall, where it evaporates, leaving bumpy deposits that grow into various shapes, resembling popcorn, grapes, potatoes, and cauliflower. A drapery is formed when water droplets flow down a sloping cave ceiling, leaving trails of calcite that build up layer by layer. Sheets of calcite, called flowstone, are deposited when wide streams of water flow down cave walls, often constructing massive terraces. Doubly terminated quartz crystals are also found in limestone caves and are sometimes cut as gemstones.

In underwater caves, such as those on the Yucatán Peninsula in Mexico, the limestone formations also include delicate, hollow stalactites called soda straws that took millions of years to create.

CAVE ART

One of the earliest human species, *Homo erectus,* appeared in Africa about 1.5 million years ago. By 1 million years ago, these people were present in southern and eastern Asia, where they lived until about 200,000 years ago. *Homo erectus* developed quite an elaborate culture, characterized by inhabiting caves and hunting game. Some populations might have been the first to use fire, possibly for hunting, cooking, and keeping warm.

People discovered fire possibly as early as 800,000 years ago, which was fortunate because without it they might not have survived the cold glacial periods in the northern latitudes. Whether humans could ignite a fire by this time is uncertain. Most likely, they used fires naturally set by lightning strikes. For thousands of years, people used fire for hunting game by setting brush fires to frighten animals into traps or off cliffs.

Peking man was a variety of *Homo erectus* that lived in a large cave about 30 miles southwest of Beijing (Peking), China. These people occupied this cave continuously for more than 200,000 years, beginning about half a million years ago. Fossilized animal bones indicated that these humans were effective hunters. Fruits and grain were also a large part of their diet, as indicated by fossilized seeds found in the cave. As early as 400,000 years ago, these people could control fire and keep it burning, although whether they could ignite it remains uncertain; they probably relied on natural blazes started by lightning strikes. Indications that they utilized fire to cook their food is evidenced by quantities of charred seeds found in the cave.

The Neandertals are generally thought to be cave dwellers because most of their bones have been found there, because the caves preserved bones better than open sites. When not occupying caves, the Neandertals lived in open-air sites, as indicated by hearths and rings made of mammoth bones and masses of stone tools normally associated with these people. The Neandertals might have even made rock carvings and cave paintings. They buried their dead and placed offerings such as ibex horns and flowers in the graves.

Neandertals also might have practiced cannibalism. In Italy's Guattari Cave, where Neandertals lived between 100,000 and 50,000 years ago, an adult male skull was found within a ring of stones from an apparent ritual of cannibalism. Human bones found at a cave in former Yugoslavia have long been viewed as the remains of a cannibal feast that occurred more than 50,000 years ago. What appears to be human cannibalism has emerged at other sites, including 60,000-year-old bones found in a cave in France. The scientific community remains divided on whether cannibalism was practiced routinely and systematically at such locations, or whether it occurred only in rare cases of imminent starvation.

Modern humans, called Cro-Magnon, originated perhaps 250,000 years ago in sub-Saharan Africa, where some of the oldest finds of our own species were discovered. However, evidence also suggests that they arose simultaneously in several parts of the world, as early as 1 million years ago, possibly evolving from *Homo erectus.* They were named for the Cro-Magnon cave in France, where the first discoveries were made in 1868. Their appearance was markedly different from the stocky Neandertals, and they shared the great majority of physical attributes of humans today.

Sometime during the last ice age, the Cro-Magnon advanced into Europe and Asia probably during a warm interlude when the climate was not so severe. Neandertals and modern humans apparently coexisted in Eurasia for at least 60,000 years and shared many of the same cultural advancements. They invented sewing needles to tailor cold-weather clothing needed for colonizing the colder regions of Europe. They buried their dead and adorned the graves with the personal belongings of their fallen kin.

Cave paintings were a common form of human expression, and one cave in the French Pyrenees has walls containing more than 200 human handprints, most with missing fingers, dating to about 26,000 years ago. The fingers might have been destroyed by disease or infection or hacked off in some sort of ritual. The late ice age people tended to make elaborate and beautiful cave drawings and carvings of animals, especially those they did not eat and held in high regard, such as horses and cave bears. The late ice age people also made cave drawings of animals they pursued in the hunt. Cave paintings in Brazil suggest that cave art began in the Americas almost at the same time it appeared in Europe and Africa.

Below sandstone cliffs such as those in the American Southwest are adobe dwellings of the Anasazi (Fig. 146), who mysteriously vanished around 800 years ago, possibly due to a prolonged period of drought. Other Indians have carved or painted a variety of figures on cave walls called petroglyphs. Originally, the carvings were thought to be simple pieces of cave art. However, many carvings bear a certain relation to the sun's path across the sky.

One of the most striking examples is called the Cave of Life, lying in the heart of the Petrified Forest in Arizona. On one stone wall is a carving of an elaborate cross. When the rays of the setting sun strike the center of the cross, it marks the shortest day of the year. Another cave at Painted Rocks near Gila Bend, Arizona, has a similar cross carved on the cavern wall probably for the same reason. Their purpose remains a mystery, however. Perhaps they were

Figure 146 *Little Long House in overhang ledge of Cliff House sandstone.*

(Photo by C. H. Dane, courtesy USGS)

used as a calendar to determine the seasons of the year so the people would know when to plant and harvest their crops.

After examining the structure of caves, the next chapter will discuss what happens when caverns and other formations collapse.

10

COLLAPSED STRUCTURES

CATASTROPHIC GROUND FAILURES

This chapter examines ground failures and collapsed structures and how they influence the shaping of the Earth's surface. Ground failures are responsible for gouging out entire sections of the crust. Landslides on unstable slopes are especially dangerous in mountainous and hilly regions. Although not as hazardous as other geologic activity, landslides are more widespread and cause considerable damage. Rockfalls are particularly spectacular, especially when involving large blocks that fall nearly vertically down a mountain face.

Slopes are the most common and among the most unstable landforms. Under favorable conditions, the ground can give way even on the gentlest slopes, contributing to the sculpture of the landscape. Slopes are therefore inherently unstable and remain only temporary features over geologic time. The weakening of sediment layers due to earthquakes can cause massive subsidence. The earth also moves when water is added to unstable sediments during heavy rains, causing various degrees of flowage. Submarine slides can be just as impressive as those on land and are responsible for much of the oceanic terrain along the outer margins of the continents.

The surface of the Earth is pockmarked by structures resulting from catastrophic collapse. This phenomenon is perhaps best demonstrated at volcanic calderas, formed when the roof of a magma chamber collapsed or when a volcano blew off its peak, leaving a broad depression. Large earthquakes whose faults cut the surface slice up the ground, producing large breaks in the crust called fissures. The dissolution of soluble materials underground or the withdrawal of fluids from subsurface sediments leads to subsidence, or horizontal depression of the surface. Other ground failures occur when subterranean sediments liquefy during earthquakes or violent volcanic eruptions, causing the ground to give way.

LANDSLIDES

Landslides are a mass movement of soil and rock material downslope under the influence of gravity, caused primarily by earthquakes and severe weather. Landslides are also induced by the removal of lateral support caused by erosion by streams, glaciers, waves, and longshore or tidal currents. They are also initiated by previous slope failures and human activity such as excavation. In addition, the ground can give way under excess loading by the weight of rain, hail, or snow.

The main types of landslides are falls, slides, and flows—either dry or wet. All slides result from the failure of earth materials under shear (plane of contact) stress. They are initiated by an increase in shear stress and a reduction of shear strength due primarily to the addition of water to a slope. The shear strength is determined by the slope geometry along with the composition, texture, and structure of the soil (Table 14). The ground might experience changes in pore pressure and water content, which acts as a lubricant between rock layers.

Particles of rock, sand, and snow dragged down a slope by gravity collide and rub against each other and the ground as they fall. With each such interaction, the particles change direction and lose energy to friction. Generally, the smaller the slope angle the less friction within the flow. The particles on the bottom in contact with the bed are slowed, while the rest of the particles glide over them in a tumbling, chaotic mass. The maximum natural inclination of a slope is called the angle of repose. It is self-regulating by triggering slides that bring the slope back to its critical state when it becomes oversteepened.

Most landslides in the United States occur in the regions of the Appalachian and Rocky Mountains and the ranges along the Pacific Coast (Fig. 147). Although individual landslides generally are not as spectacular as other violent forms of nature, they are more widespread and can cause major economic losses and casualties. The direct costs arising from damage to highways, buildings, and other facilities and indirect costs resulting from loss of productivity can amount to more than a billion dollars annually. Single large

TABLE 14 SUMMARY OF SOIL TYPES

Climate	Temperate (Humid) > 160 in. Rainfall	Temperate (Dry) < 160 in. Rainfall	Tropical (Heavy Rainfall)	Arctic or Desert
Vegetation	Forest	Grass and brush	Grass and trees	Almost none, no humus development
Typical area	Eastern U.S.	Western U.S.		
Soil type	Pedalfer	Pedocal	Laterite	
Topsoil	Sandy, light-colored; acid	Enriched in calcite; white color	Enriched in iron and aluminum, brick red color	No real soil forms because no organic material. Chemical weathering very low
Subsoil	Enriched in aluminum, iron, and clay; brown color	Enriched in calcite; white color	All other elements removed by leaching	
Remarks	Extreme development in conifer forest abundant humus makes groundwater acid. Soil light gray due to lack of iron	Caliche—name applied to accumulation of calcite	Apparently bacteria destroy humus, no acid available to remove iron	

landslides can run up damage bills in the tens of millions of dollars. Fortunately, landslides have not resulted in a major loss of life as in other parts of the world because most catastrophic slope failures in the United States generally take place in sparsely populated areas.

The majority of landslides occur during earthquakes. A spectacular landslide triggered by the 1959 Hebgen Lake, Montana earthquake moved from north to south, gouging out a large scar in the mountainside (Fig. 148). The debris traveled uphill on the south side of the valley and dammed the Madison River, creating a large lake. The 1971 San Fernando, California earthquake unleashed nearly 1,000 landslides, distributed over a 100-square-mile area of remote and hilly mountainous terrain. During the 1976 Guatemala City earthquake, some 10,000 slides were triggered throughout an area of 6,000 square miles.

The greatest earthquakes to strike the continental United States in recorded history took place near New Madrid, in southeastern Missouri on

Figure 147 *A rockslide that blocked Highway 1, south of Big Sur, California, during severe winter storms in 1982–83.*

(Photo by G. F. Wieczorek, courtesy USGS)

the banks of the Mississippi River. During the winter of 1811–12, three massive earthquakes with estimated magnitudes ranging upward of 8.7 struck the region. The town itself was demolished when the ground beneath it collapsed 12 feet from a height of 25 feet above the river. Deep fissures opened in the

Figure 148 *The August 1959 Madison Canyon slide, Madison County, Montana.*

(Photo by J. R. Stacy, courtesy USGS)

earth, and the ground slid down from bluffs and low hills. Thousands of broken trees fell into the river, and whole sandbars and islands disappeared. The earthquake changed the course of the Mississippi, and created large lakes in the basins of down-dropped crust, the largest of which is the 50-foot-deep Reelfoot Lake (Fig. 149).

California is well known for its large earthquakes. Fortunately, most have taken place in scarcely populated regions. The hamlet of Lone Pine in the Owens Valley, east of the Sierra Nevada, was destroyed on March 26, 1872 by one of the largest earthquakes in California's history. It opened a deep fissure along a 100-mile line in the Owens Valley. At least 30 people died when their fragile adobe huts collapsed on them. More than 1,000 aftershocks ran through the area during the next three days.

During the 1906 San Francisco earthquake, the ground subsided under buildings, causing them to collapse. Landslides struck in many places, and an entire hillside slid down a shallow valley for a distance of half a mile. South of

Cape Fortunas a hill slid bodily into the sea and created a new cape. Roads and fences crossing the fault were offset horizontally up to 21 feet. Trees were uprooted, and sand boils and fissures appeared in many districts completely changing the landscape (Fig. 150).

The March 27, 1964 Alaskan earthquake, which devastated Anchorage and other nearby seaports, was the largest ever recorded on the North American continent. The earthquake set off landslides, and 30 blocks of Anchorage were destroyed when the city's slippery clay substratum slid toward the sea. Huge fissures opened in the outlying areas, and some of the greatest crustal deformation ever known took place (Fig. 151). The area of destruction was estimated at 50,000 square miles, and the earthquake was felt over an area of 500,000 square miles.

In volcanic regions, seismic activity, uplift, and the presence of thick deposits of unconsolidated pyroclastic material create ideal conditions for landslides. The distribution of landslides is controlled by the seismic intensity, topographic amplification of the ground motion, lithology (rock type), slope steepness, and regional fractures or other weaknesses in the rock. Heavy, sustained rainfall over a wide area also can trigger landslides in volcanic terrain.

During the 1980 explosive eruption of Mount St. Helens, a wall of earth slid down the mountainside, creating one of the greatest landslides in modern

Figure 149 *Reelfoot Lake, Tennessee, created by flooding of down-dropped crust from the 1811–12 New Madrid, Missouri, earthquake.*

(Photo by Fuller, courtesy USGS)

Figure 150 *Fissures along the San Andreas Fault between Point Reyes Station and Olema, Marin County, California, from the 1906 San Francisco earthquake.*

(Photo by G. K. Gilbert, courtesy USGS)

times. It filled the valley below with debris covering 20 square miles. One arm of the gigantic mass plowed through Spirit Lake at the base of the volcano and burst into the valley beyond, devastating everything in its path for 18 miles (Fig. 152). Massive mudflows scoured the slopes of the volcano and jammed

Figure 151 *Wreckage of Government Hill School due to catastrophic subsidence from the March 27, 1964, Alaskan earthquake, Anchorage District, Alaska.*

(Photo by G. K. Gilbert, courtesy USGS)

the Cowlitz and Columbia Rivers to the Pacific Ocean with debris and timber blown down by the tremendous blast.

Large parts of Alaska's Mount St. Augustine have collapsed and fallen into the sea, generating large tsunamis. Massive landslides have ripped out the flanks of the volcano 10 or more times during the past 2,000 years. The most recent slide coincided with the October 6, 1883 eruption, when debris on the flanks of the volcano crashed into the Cook Inlet, sending a 30-foot tsunami to Port Graham 54 miles away that destroyed boats and flooded houses. Subsequent eruptions have filled the gap left by the last landslide, making the volcano increasingly unstable and ready for another major collapse. A recurring landslide would barrel down the north side of the volcano and plunge into the sea, sending a tsunami in the direction of cities and oil platforms residing in the inlet.

On the south flank of Kilauea Volcano on the southeast coast of Hawaii, about 1,200 cubic miles of rock are moving toward the sea at speeds of up to 10 inches per year. The earth movement is presently the largest on the planet and could ultimately lead to catastrophic sliding comparable to those of the past that have left massive piles of rubble on the ocean floor. Slides play an important role in building up the continental slope and the deep abyssal plains, making the seafloor one of the most geologically active places on Earth.

Figure 152 *Destruction from the 1980 eruption of Mount St. Helens, showing mud-clogged Spirit Lake in the foreground.*

(Photo by R.V. Emetaz, courtesy USDA Forest Service)

Rockslides are usually large and destructive, involving millions of tons of rock. They develop when planes of weakness, such as bedding planes or jointing, are parallel to a slope, especially if the slope has been undercut by a river, a glacier, or construction work. When a mass of bedrock is broken into many fragments during the fall, it behaves as a fluid that spreads out in the valley below. It might even flow some distance uphill on the opposite side of the valley. Such slides are commonly called avalanches, but this term is generally applied to snowslides (Fig. 153).

Figure 153 *A slab avalanche triggered by a skier.*

(Photo courtesy USGS)

One of the worst avalanches in recent history occurred in the foothills of the Himalayas in Kashmir, northern India, on January 16, 1995. A blizzard had stranded hundreds of people who abandoned their cars and buses on a one-lane highway to take shelter inside a 1.5-mile-long tunnel. Without warning, a large avalanche struck, burying everything in the area. Some people managed to escape from the tunnel before thousands of tons of snow completely closed it off. Several days later, bulldozers and villagers armed with shovels dug through the wall of snow, only to find the tunnel filled with frozen bodies.

During one of the most destructive rockslides, the town of Elm, Switzerland, was wiped off the map on September 11, 1881, when a nearby mountainside collapsed, transforming a solid cliff into a river of rock. The cliff-side plummeted 2,000 feet and sped through the valley below covering a distance of nearly 1.5 miles. As the gigantic mass of rock debris roared down the valley, it entombed 116 people beneath a thick blanket of broken slate before grinding to a halt.

Material that drops from a near vertical mountain face is called a rock-fall or soilfall. Rockfalls can range in size from individual blocks plunging down a mountain slope to the failure of huge masses, weighing hundreds of thousands of tons falling nearly straight down a mountain face. Individual blocks commonly come to rest in a loose pile of angular blocks at the base of a cliff, called talus. If large blocks of rock drop into a standing body of water, immensely destructive waves are set in motion. A 1958 earthquake in Alaska produced an enormous rockslide that fell into Lituya Bay and generated a wave of water surging 1,720 feet up the mountainside. Trees were bowled over, and the shores along the bay were inundated with water that wiped out everything in its path (Fig. 154).

The most celebrated example of a rockfall in North America took place in Alberta, Canada, in 1903. A mass of strongly jointed limestone blocks at the crest of Turtle Mountain, possibly undercut by coal mining carried on below the base, broke loose and plunged down the deep escarpment. Some 40 million tons of material fell down the mountainside and washed through the small coal-mining town of Frank in one gigantic wave, killing 70 people along the way. The rock-fall then swept up the opposite slope 400 feet above the valley floor.

A rockfall southeast of Glacier Point in Yosemite National Park, California, on July 10, 1996 sent 160,000 tons of granite that broke off a cliff, plunging a third of a mile at more than 160 miles per hour. The slide caused a hurricane-like "air blast" that leveled thousands of trees, some with their bark completely stripped off. The air blast represents a poorly understood collateral hazard of rockfalls similar to dropping a book parallel to the ground, which forces the air out from underneath it. As a result, geologists might have to reassess hazard zones marked on maps at Yosemite and other mountainous national parks to take into account the danger from air blasts.

Figure 154 *Wave damage on the south shore of Lituya Bay, Alaska, from a massive rock slide in 1958.*

(Photo by D. J. Miller, courtesy USGS)

Other forms of earth movements are slumps, which develop when a strong, resistant rock overlies weaker strata. Material slides down in a curved plane, tilting up the resistant unit, while the weaker rock flows out to form a heap. Unlike rockslides, slumps develop new cliffs just below those previous to the slump, setting the stage for renewed slumping. Thus, slumping is a continuous process, and generally, many generations of slumps lie far in front of the present cliffs.

When loosened by rain or melting snow, ordinary soil on a steep hillside can suddenly turn into a wave of sediment, sweeping downward at speeds of more than 30 miles per hour. Precipitation can free dirt and rocks by increasing the water pressure inside pores within the soil. As the water table rises and pore pressure increases, friction holding the top layer of soil to the hillside begins to drop until the pull of gravity overcomes it. Immediately before the soil begins to slide, the pore pressure drops, which signals that the soil is beginning to expand just before it starts to slide.

Soil slides occur in weakly cemented fine-grained materials that form steep stable slopes under normal conditions, but fail during earthquakes. The size of the area affected by earthquake-induced landslides depends on the magnitude and focal depth of the earthquake, the topography and geology of the ground near the fault, and the amplitude and duration of the temblor. Soil-flow failures as much as 1 mile in length and breadth caused by the great 1920 Kansu (Gansu),

China, earthquake killed an estimated 180,000 people. As the temblor rumbled through the region, immense slides rushed out of the hills, burying entire villages, damming streams, and turning valleys into instant lakes.

Soils and soft rocks that swell or shrink due to changes in moisture content are called expansive soils. Damage to buildings and to other structures built on expansive soils can be very costly. The soils are abundant in geologic formations in the Rocky Mountain region, the Basin and Range Province, most of the Great Plains, much of Gulf Coastal Plain, the lower Mississippi River Valley, and the Pacific Coast. The parent materials for expansive soils are derived from volcanic and sedimentary rocks that decompose to form expansive clay minerals such as montmorillonite and bentonite. These materials are often used as drilling mud because of their ability to absorb large quantities of water. Unfortunately, this characteristic also causes them to form highly unstable slopes.

LIQUEFACTION

Ground failures during earthquakes and violent volcanic eruptions, resulting from the failure of subterranean sediments saturated with water, are caused by liquefaction. Earthquakes can turn a solid, water-saturated bed of sand underlying less permeable surface layers into a pool of pressurized liquid that rises to the surface, sometimes causing localized flooding. Generally, the younger and looser the sediment and the shallower the water the more susceptible the soil is to liquefaction.

Liquefaction causes clay-free soils, primarily sands and silts, to temporarily lose strength and behave as viscous fluids rather than as solid materials. It occurs when seismic shear waves pass through a saturated granular soil layer, which distorts its structure and causes some void spaces to collapse in loosely packed sediments. Each collapse transfers stress to the pore water surrounding the grains. Disruptions to the soil generated by these collapses increases pressure in the pore water, causing it to drain. If the drainage is restricted, the pore-water pressure builds. When the pore-water pressure reaches the pressure exerted by the weight of the overlying soil, grain contact stress is temporarily lost and the granular soil layer flows fluidlike.

The three types of ground failure associated with liquefaction are lateral spreads, flow failures, and loss of bearing strength. Lateral spreads are the lateral movement of large blocks of soil due to liquefaction in a subsurface layer caused by earthquakes. They generally develop on gentle slopes of less than 6 percent angle. Horizontal movements on lateral spreads can extend up to 15 feet, but where slopes are particularly favorable and the duration of the temblor is long, lateral movement might range up to 10 times farther. Lateral spreads usually break up internally, forming scarps and fissures (Fig. 155).

Figure 155 *Ground fractures in Forest Acres area from lateral spreading during the March 27, 1964, Alaskan earthquake, Seward District, Alaska Gulf region.*

(Photo by C. D. Miller, courtesy USGS)

During the 1964 Alaskan earthquake, more than 200 bridges were damaged or destroyed by lateral spreading of floodplain deposits near river channels. These spreading deposits compressed bridges over the channels, buckled decks, thrust sedimentary beds over abutments, and shifted and tilted abutments and piers. Lateral spreads are also destructive to pipelines. During the 1906 San Francisco earthquake, major water main breaks occurred, which hampered firefighting efforts. The inconspicuous ground failure displacements, some as much as 7 feet, were largely responsible for the destruction of San Francisco (Fig. 156).

Figure 156 *Secondary cracks in pavement from the 1906 San Francisco earthquake in California.*

(Photo by G. K. Gilbert, courtesy USGS)

Flow failures are the most catastrophic type of ground failure due to liquefaction. They consist of liquefied soil or blocks of intact material riding on a layer of liquefied soil. Flow failures usually move dozens of feet, but under certain geographic conditions they can travel several miles at speeds of many miles per hour. They commonly form in loose saturated sands or silts on slopes greater than 6 percent angle and originate both on land and on the seafloor.

Most clays lose strength when disturbed by earthquakes. If the loss of strength is large, some clays, called quick clays, might fail. Quick clay is composed primarily of flakes of clay minerals arranged in very fine layers, with a water content often exceeding 50 percent. Ordinarily, quick clay is a solid that can support over a ton per square foot of surface area. However, the slightest jarring motion from an earthquake can immediately cause it to liquefy.

The 1964 Alaskan earthquake initiated landslides and ground subsidence that were highly destructive. The ground beneath Valdez and Seward literally gave way, and both waterfronts floated toward the sea. In Anchorage, houses were destroyed when 200 acres were carried toward the ocean. The five large landslides that affected parts of Anchorage are examples of spectacular failures of clays sensitive to ground motions. The slides resulted from the failure of layers of quick clay along with other layers composed of saturated sand and silt. The severity of the earthquake caused a loss of strength in the clay layers and liquefaction in the sand and silt layers. These were the major contributing factors to the landsliding and subsidence that destroyed a major portion of the city (Fig. 157).

Figure 157 *The Fourth Avenue slide area in Anchorage from the March 27, 1964 Alaskan earthquake.*

(Courtesy USGS and U.S. Army)

Figure 158 *Apartment buildings in Niigata, Japan, that tipped because of the loss of bearing strength caused by liquefaction in the underlying sediments.*

(Courtesy of USGS)

The largest and most damaging flow failures have taken place undersea in coastal areas. Submarine flow failures carried away large sections of the port facilities at Seward, Whittier, and Valdez, Alaska, during the 1964 earthquake. Submarine flow failures also can generate large tsunamis that overrun parts of the coast. For example, on July 3, 1992 what appeared to be a large undersea slide sent a 25-mile-long, 18-foot-high wave crashing down on Daytona Beach, Florida, overturning automobiles and injuring 75 people. In 1929, an earthquake on the coast of Newfoundland set off a large undersea landslide and triggered a tsunami that killed 27 people.

When the soil supporting buildings or other structures liquefies and loses strength, large deformations can occur within the soil, causing buildings to settle or tip over. Soils that liquefy beneath buildings distort the subsurface geometry, causing bearing failures and subsequent subsidence that can tilt buildings. Normally, these deformations occur when a layer of saturated, cohesionless sand extends from near the surface to a depth about equal to the width of the building. The most spectacular example of this type of ground failure occurred during the June 16, 1964 Niigata, Japan, earthquake, when several four-story apartment buildings tilted as much as 60 degrees (Fig. 158).

The earthquake caused sections of the city to subside a foot or more, resulting in serious flooding when dikes holding back the sea were breached.

MASS WASTING

Not all Earth movements are induced by earthquakes. Many are caused by mass wasting, which is the mass transfer of material downslope by the direct influence of gravity. Mass wasting causes slipping, sliding, and creeping even down the gentlest slopes. Creep (Fig. 159) is the slow downslope movement of bedrock and overburden, the soil overlying bedrock. It is recognized by downhill tilted poles and fence posts, indicating a more rapid movement of near surface soil material than that below.

Normally, trees are unable to root themselves, and only grass and shrubs grow on the slope. Sometimes where the creep is slow, the trunks of

Figure 159 *Railroad tracks damaged by soil creep near Coal Creek, Canada.*

(Photo by W. W. Atwood, courtesy USGS)

trees are bent, and after the trees are tilted, new growth attempts to straighten them. However, if the creep is continuous, the trees lean downhill in their lower parts and become progressively straighter higher up. Creep can be very rapid where frost action is prominent. After a freeze-thaw sequence, material moves downslope due to the expansion and contraction of the ground.

A rise in the water content of the overburden increases the weight and reduces stability by lowering resistance to shear, resulting in an earthflow, a more visible form of movement. Earthflows are characterized by grass-covered, soil-blanketed hills. Although generally minor features, some can be quite large, covering several acres. Earthflows usually have a spoon-shaped sliding surface, whereupon a tongue of overburden breaks away and flows for a short distance. An earthflow differs from creep in that a distinct, curved scarp is formed at the breakaway point.

With a further increase in water content, an earthflow might grade into a mudflow (Fig. 160). The behavior of mudflows is similar to that of a viscous fluid, often carrying a tumbling mass of rocks and large boulders. They are also produced by rain falling on loose pyroclastic material (rocks formed by volcanic explosion) on the flanks of certain types of inactive volcanoes. Mudflows are the most impressive feature of the world's deserts. Heavy runoff forms rapidly moving sheets of water that pick up huge quantities of loose material. The floodwaters flow into the main stream, where all the muddy material is suddenly concentrated in the main channel. The dry streambed is rapidly transformed into a flash flood that moves swiftly downhill, in some cases with a steep, wall-like front.

This type of mudflow can cause considerable damage as it flows out of mountain ranges. Eventually, the loss of water by percolation into the ground thickens the mudflow until it ceases to flow. Mudflows often carry large blocks and boulders onto the floor of the desert basins far beyond the base of the bordering mountain range. Huge monoliths rafted out beyond the mountains by swift-flowing mudflows are left standing in the middle of nowhere (Fig. 161).

Mudflows arising from volcanic eruptions are called lahars, from the Indonesian word meaning "mudflow," named so because of their large occurrence in this region. Lahars are masses of water-saturated rock debris that move down the steep slopes of a volcano, resembling the flowage of wet concrete. The debris is commonly derived from masses of loose unstable rock deposited on the volcano by explosive eruptions. The water is provided by rain, melting snow, a crater lake, or a reservoir next to the volcano. Lahars are also initiated by a pyroclastic or lava flow moving across a snowfield, rapidly melting it. They can be either cold or hot, depending on whether hot rock debris is present.

Figure 160 *The Slumgullion mudflow, Himsdale County, Colorado in 1905.*

(Photo by W. Cross, courtesy USGS)

One of the world's worst volcanic mudflows in recent history took place on November 13, 1985 when the eruption of Nevado del Ruiz Volcano in Colombia melted the mountain's icecap and sent floods and mudflows cascading 30 miles per hour down its sides. A 130-foot wall of mud and ash careened down the narrow canyon. When it reached the town of Armero 30 miles away, the mudflow spread out and flowed rapidly through the city streets, creating 10-foot-high waves. The deluge buried most of the town and other nearby villages, killing more than 25,000 people.

The speed of lahars depends mostly on their fluidity and the slope of the terrain. They can move swiftly down valley floors for a distance of up to 50 miles or more at speeds exceeding 20 miles per hour. Lava flows that extend into areas of snow or glacial ice might melt them, producing floods as well as lahars. Flood-hazard zones extend considerable distances down some valleys.

Figure 161 *A 700-ton boulder transported by the May 31, 1970, Peruvian earthquake.*

(Photo courtesy USGS)

For the volcanoes in the western Cascade Range, these zones can reach as far as the Pacific Ocean.

The most common triggering mechanisms for mass wasting include vibrations from earthquakes or explosions that break the bond holding the slope together, overloading the slope so it can no longer support its new weight, undercutting at the base of the slope, and oversaturating the slope with water. Water adds to the weight of the slope and lessens the internal cohesion of the overburden. Although the effect of water as a lubricant is commonly considered its main role, this function is actually quite limited. The main effect is the loss of the cohesion of the material by filling the spaces between soil grains with water.

Another type of movement of soil material is called frost heaving. It is associated with cycles of freezing and thawing mainly in the temperate climates. Frost heaving thrusts boulders upward through the soil by a pull from above and by a push from below. If the top of the rock freezes first, it is pulled upward by the expanding frozen soil. When the soil thaws, sediment gathers below the rock, and it settles at a slightly higher level. The expanding frozen soil lying below also heaves the rock upward. After several frost-thaw cycles, the boulder finally comes to rest on the surface, a major annoyance to northern farmers, who find a new crop of rocks in their fields every spring. Rocks have also been known to push through highway pavement, and fence posts have been shoved completely out of the ground.

Frost action can produce mechanical weathering by exerting pressures against the sides of cracks and crevices in rocks when water freezes inside them, resulting in frost wedging. This widens the cracks, while surface weathering rounds off the edges and corners, providing a landscape resembling multitudes of miniature canyons up to several feet wide carved into solid bedrock.

SUBSIDENCE

Subsidence is the lowering or collapse of the land surface either locally or over broad regional areas by the withdrawal of fluids and by vibrations from earthquakes. Because underground fluids fill intergranular spaces and support sediment grains, the removal of large volumes of fluid such as water or oil results in a loss of grain support, a reduction of intergranular void spaces, and the compaction of clays. This causes subsurface compaction and subsequent land surface subsidence.

Many parts of the world have been steadily sinking due to the withdrawal of large quantities groundwater or petroleum. In the United States, the most dramatic examples of subsidence occur along the gulf coast of Texas, in Arizona, and in California. Large areas of California's San Joaquin Valley have subsided because of intensive pumping of groundwater. The arid agricultural region is so dependent on groundwater it accounts for up to 20 percent of all well water pumped in the United States. The ground is sinking at rates of up to a foot per year, and in some places the land has fallen more than 20 feet below former levels.

The Houston-Galveston area of Texas has experienced local subsidence as much as 7.5 feet and more than 1 foot over an area of 2,500 square miles, mostly due to the withdrawal of large amounts of groundwater. In Galveston Bay, subsidence of 3 feet or more occurred over an area of several square miles due to the rapid pumping of petroleum from the underlying strata. Some coastal towns have subsided such that they are susceptible to flooding during hurricanes. In Mexico City, overpumping of groundwater has caused some parts of the city to subside at a rate of more than a foot per year, often resulting in earth tremors. This might explain why residents ignored the foreshocks that preceded the destructive September 19, 1985 earthquake of 8.1 magnitude that destroyed a large portion of the city.

One of the most dramatic cases of subsidence caused by the withdrawal of petroleum occurred at Long Beach, California, where the ground sank to form a huge bowl as much as 26 feet deep over an area of 22 square miles. The affected land subsided at the rate of 2 feet per year in some parts of the oil field, and in the downtown area the subsidence amounted to nearly 6 feet, causing considerable damage to the city's infrastructure. Most of the subsidence was halted by injecting large quantities of seawater under high pressure back into the underground reservoir. The seawater injection also produced a secondary benefit by increasing the production of the oil field, because oil floats on water and rises toward the surface.

Venice, Italy, is slowly drowning because the sea is going up while the city is going down. Much of the subsidence is due to overuse of groundwater, causing the aquifer (water-filled strata) under the city to compact. The cumulative subsidence ofVenice has been slightly more than 5 inches over the latter half of the 20th century. Meanwhile, the Adriatic Sea has risen 3.5 inches due to an apparent global warming, resulting in a thermal expansion of the ocean and the melting of the ice caps. Together with the subsidence, that makes for a change of more than 8 inches between Venice and the sea.

Port Said along the northeast coast of Egypt's Nile Delta is a bustling seaport of half a million people. The region overlies a large depression filled with between 40 and 160 feet of mud, indicating that part of the delta is slowly dropping into the ocean. Over the last 8,500 years, this portion of the fan-shaped delta has been lowering by less than a quarter inch per year, but more recently the yearly combined subsidence and sea level rise has exceeded this amount. Presently, the delta is just 3 feet above sea level, and a 2- to 3-foot expected rise in sea level by the end of the 21st century could place major portions of the city underwater. Moreover, as the land subsides, seawater infiltrates into the groundwater system, rendering it useless. Dams and human-made channels have nearly completely cut off the region's sediment supply from upriver, preventing eroded areas along the ocean from building back up.

Coastal subsidence induced by earthquakes causes vegetated lowlands that are sufficiently elevated to avoid being inundated by the sea to sink far enough to be submerged regularly and become barren tidal mud flats. Between great earthquakes, sediments fill the tidal flats and raise them to the level where vegetation can again grow. Therefore, repeated earthquakes produce alternating layers of lowland soil and tidal flat mud.

Subsidence due to the withdrawal of groundwater can produce fissures or the renewal of surface movement in areas cut by preexisting faults. The fissuring results in the formation of open cracks in the ground. Surface faulting and fissuring resulting from the withdrawal of groundwater is a potential problem near Las Vegas, Nevada, as well as parts of Arizona, California, Texas, and New Mexico. The withdrawal of large volumes of water along with oil and gas also can cause the ground to subside to considerable depths, sometimes with catastrophic consequences.

Subsidence caused by earthquakes in the United States has taken place mainly in Alaska, California, and Hawaii. The subsidence results from the vertical movement on faults that can affect broad areas. During the 1964 Alaskan earthquake, some 70,000 square miles were tilted downward more than 3 feet, causing extensive flooding. Intense earthquakes can cause subsidence over smaller areas. During the 1811–12 New Madrid, Missouri earthquakes, subsurface sand and water were ejected to the surface, leaving underground voids that caused compaction of subterranean materials and ground settling.

During an earthquake, sand boils often develop in young sedimentary environments where the water table is near the surface. Sand boils are fountains of water and sediment that spout from the pressurized liquefied zone and can reach up to 100 feet high. They are produced when water laden with sediment is vented to the surface by artesianlike water pressures that can excavate large sand pits (Fig. 162). Sand boils also can cause local flooding and the accumulation of huge amounts of silt and sand, often in places where they are not wanted. The expulsion of sediment-laden fluids from below ground might also form a large subsurface cavity that causes the overlying layers to subside.

The settling of sediments after the addition of water causes significant subsidence, especially in the dry western states that have been heavily irrigated, such as the San Joaquin Valley of California. Subsidence occurs when dry surface or subsurface deposits are extensively wetted for the first time since their deposition. The wetting causes a reduction in the cohesion between sediment grains, allowing them to move and fill intergranular openings. This results in the lowering of the land surface from 3 to 6 feet and as much as 15 feet in the most extreme cases. The effects of the compaction on the land are usually uneven, causing depressions, cracks, and wavy surfaces.

CATASTROPHIC COLLAPSE

Another type of ground failure that occurs in colder climates is called solifluction. When frozen ground melts from the top down, as during warm spring days in the temperate regions or during the summer in permafrost areas, it causes soil to move downslope over a frozen base. Buildings are seriously damaged by the loss of foundation support as they are being carried away.

The Nevada Test Site, 65 miles northwest of Las Vegas, has taken on the appearance of a moonscape, pockmarked with craters created by underground nuclear tests. The craters formed when subterranean sediments fused into glass by the tremendous heat generated by the explosions. This greatly reduced the volume, causing the overlying sediments to collapse to fill the underground

Figure 162 *Sandblows from the October 15, 1979, Imperial Valley, California, earthquake.*

(Photo by C. E. Johnson, courtesy USGS)

caverns. Sometimes fissures opened on the surface to vent gases escaping from the molten rocks.

Collapse of abandoned underground mines, especially shallow coal mines, occurs often in the eastern United States. The rocks above the mine workings

Figure 163 *Subsidence depressions, pits, and cracks above abandoned underground mine, Sheridan County, Wyoming.*

(Photo by C. R. Dunrud, courtesy USGS)

might not have adequate support. When they collapse, the surface drops several feet, forming many depressions and pits (Fig. 163). Solution mining, using water to remove soluble minerals such as salt, gypsum, and potash, can produce huge underground cavities, whose collapse causes surface subsidence.

The forgotten shafts left in old coal and salt mines might collapse under overlying buildings if not backfilled. However, locating the mines can often be difficult. Because an underground cavity has a higher electrical resistance than the surrounding materials, its location and dimensions can be determined by monitoring electrical fields that travel into the ground. The technique also might be useful for detecting underground tunnels and caverns for archaeologists looking for buried artifacts.

After a discussion of catastrophic collapse, the next chapter will examine another type of crater caused by meteorite impacts.

11

METEORITE IMPACT CRATERS

ASTEROID AND COMET COLLISIONS

This chapter investigates asteroids and comets and their impacts on the Earth. Throughout its long history, the Earth has been repeatedly bombarded by asteroids and comets, with a much higher incidence during the early years than more recently. This is fortunate because the evolution of life would have turned out much differently if the impact rate had remained high. In its early development, the Earth was heavily bombarded and possibly struck by as many as three Mars-size bodies, one of which might have created the moon.

Sometimes asteroids the size of mountains struck the planet, inflicting a great deal of damage and causing the extinction of species. Massive comet swarms, involving perhaps thousands of comets impacting all over the Earth, might also explain the disappearance of species. A popular theory for the extinction of the dinosaurs contends that the planet was hit by a large asteroid or comet nucleus that excavated a deep crater 100 miles or more wide and caused ecologic chaos.

The search on land and sea for ancient impact craters is difficult because of the Earth's highly active geology, which has long since erased all but the faintest signs. Impacts on the moon and the inner planets as well as the moons of the outer planets are quite evident and numerous. The Earth was probably hit by a dozen times as many meteorites as its moon because of the planet's larger size and greater gravitational attraction. However, the moon retains a better record of terrestrial impact cratering.

Fortunately, several remnants of ancient terrestrial craters remain, suggesting the Earth was just as heavily bombarded as the rest of the solar system. Many strikingly circular features have been found that appear to be impact craters. Due to their low profiles and subtle stratigraphy, however, they were previously unrecognized as impact structures. In the future, many more craters are bound to be discovered by sophisticated instruments aboard satellites, providing a clearer picture of what had transpired long ago.

THE ASTEROID BELT

Between the orbits of Mars and Jupiter lies the asteroid belt, comprising about a million pieces of solar system rubble larger than a mile across, along with more numerous smaller objects. Zodiacal dust bands consisting of fine material orbit the sun near the inner edge of the main asteroid belt (Fig. 164). The debris is believed to have originated from comet trails, composed of dust and gas blown outward by the solar wind, and from collisions between asteroids.

Not all asteroids reside within the main belt. An interesting group called Trojans lies in the same orbit as Jupiter. Another asteroidlike body has a far-ranging orbit that carries it from near Mars to beyond Uranus. One large object called Chiron lies between the orbits of Saturn and Uranus, indeed an odd place for an asteroid.

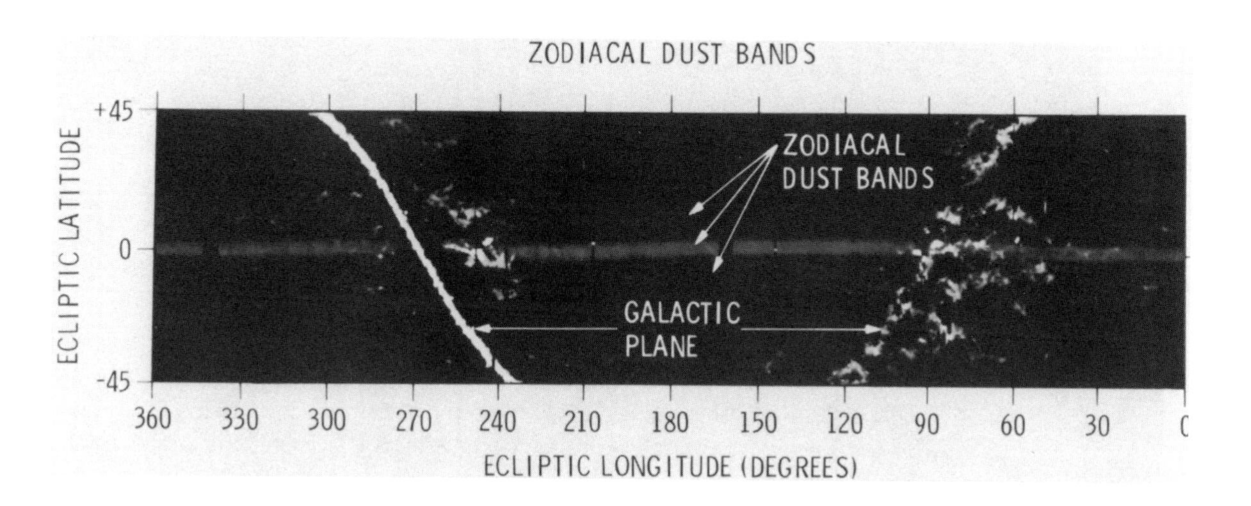

Figure 164 *The zodiacal dust bands are debris from comets and collisions between asteroids near the inner edge of the asteroid belt.*

(Photo courtesy NASA)

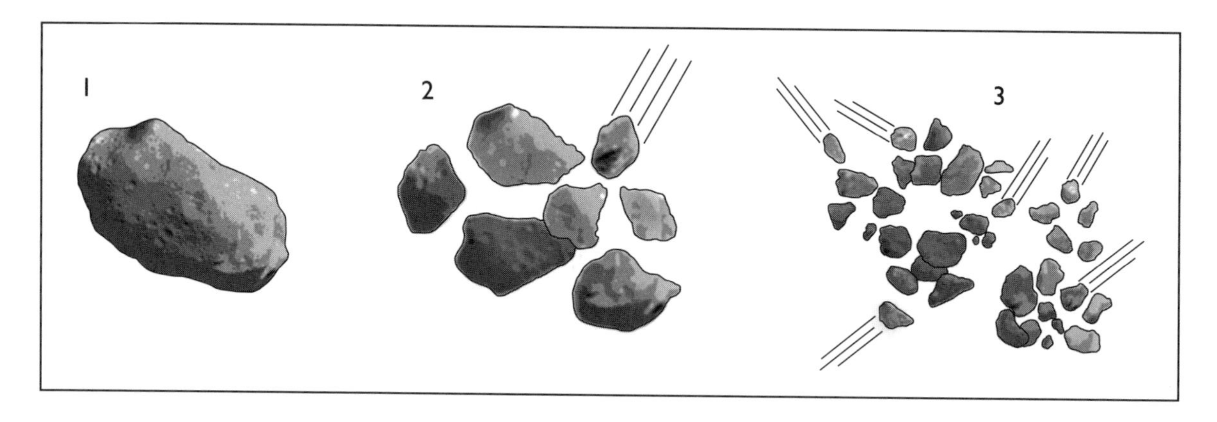

Figure 165 *(1) A planetoid smaller than the Moon is (2) broken up by a giant collision, and (3) additional collisions yield asteroids that bombard the Earth.*

Asteroids are leftovers from the creation of the solar system, and due to the strong gravitational attraction of Jupiter they were unable to coalesce into a single planet. Instead, they formed several planetoids smaller than the moon as well as a broad band of debris, called meteoroids, which are fragments broken off asteroids by numerous collisions. Originally, the combined masses of all the material in the asteroid belt was nearly equal to the present mass of the Earth. However, constant collisions have weeded out the asteroids, so that now their combined mass is perhaps less than 1 percent of the original.

A large portion of the asteroids contain a high concentration of iron and nickel, suggesting they were once part of the metallic core of a planetoid that disintegrated after a collision with another body. Some large asteroids might have melted and differentiated early in the formation of the solar system. The inner and middle belt asteroids underwent a great deal of heating and experienced as much melting as did the planets. The molten metal in the asteroids along with siderophiles (iron lovers), such as iridium and osmium of the platinum group, sank to their interiors and solidified. The metallic cores were exposed after eons of collisions between asteroids chipped away the more fragile surface rock. Thus, breakup after collisions yielded several dense, solid fragments (Fig. 165).

The stony asteroids, which are much less dense and contain a high percentage of silica, exist near the inner part of the asteroid belt. The darker carbonaceous asteroids, which contain a high percentage of carbon, lie toward the outer portion of the asteroid belt. Between these regions are wide spaces called Kirkwood gaps, named for the American mathematician Daniel Kirkwood, that are almost totally devoid of asteroids. If an asteroid falls into one of these gaps, its orbit stretches, causing it to swing in and out of the asteroid belt and bringing it close to the sun and the orbits of the inner planets.

ASTEROIDS AND COMETS

Asteroids are a relatively recent discovery. On January 1, 1801, while searching for the so-called "missing planet" in the wide region between Mars and Jupiter, the Italian astronomer Giuseppe Piazzi discovered instead the asteroid Ceres, named for the guardian goddess of Sicily. It is the largest of the known asteroids, with a diameter of more than 600 miles (Table 15).

Asteroids, from the Greek meaning "starlike," and once thought to be the debris from a shattered Mars-sized planet, are actually the remnants of a planet that failed to form. Therefore, asteroids offer significant evidence for the creation of planets and provide clues to conditions in the early solar system.

Asteroids and comets are distinctly different inhabitants of the solar system. Surrounding the Sun about a light-year away is a shell of more than a trillion comets with a combined mass of 25 Earths, called the Oort Cloud, after the Dutch astronomer Jan Kendrick Oort. Another band of comets exists closer to the Sun, called the Kuiper belt comets, but they are still well beyond Pluto, which due to its odd orbit might itself be a captured comet nucleus or an asteroid.

Comets, such as Comet Halley (Fig. 166), which reappeared in 1985–86 after spending 76 years in deep space, are hybrid planetary bodies consisting of a stony inner core and an icy outer layer. Comets are characterized as flying icebergs mixed with small amounts of rock debris, dust, and organic matter. They are believed to be aggregates of tiny mineral fragments coated with organic compounds and ices enriched in the volatile elements hydrogen,

TABLE 15 SUMMARY OF MAJOR ASTEROIDS

Asteroid	Diameter (Miles)	Distance from Sun (Million Miles)	Type
Ceres	635	260	carbon-rich
Pallas	360	258	rocky
Vesta	344	220	rocky
Hygeia	275	292	carbon-rich
Interamnia	210	285	rocky
Davida	208	296	carbon-rich
Chiron	198	1270	carbon-rich
Hektor	185X95	480	uncertain
Diomedes	118	472	carbon-rich

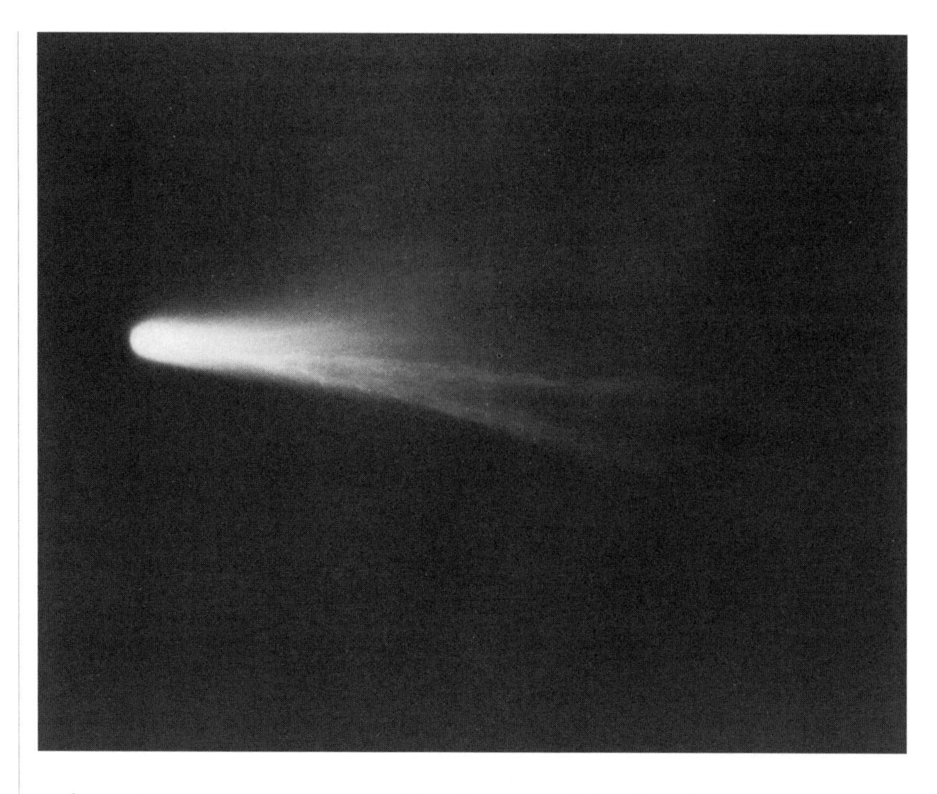

Figure 166 *Comet Halley, viewed from the National Optical Astronomy Observatories.* (Photo courtesy NOAO)

carbon, nitrogen, oxygen, and sulfur. Comets might therefore more accurately be described as frozen mudballs with equal volumes of ice and rock.

Most comets travel around the sun in highly elliptical orbits that carry them a thousand times farther out than the planets. Only when they swing close by the sun, traveling at fantastic speeds, do the ices become active and outgas large amounts of matter. As the comet journeys into the inner solar system, carbon monoxide ice vaporizes first and is replaced by jets of water vapor as the driving force behind the comet's growing brightness. Water vapor and gases stream outward, forming a tail millions of miles long that points away from the sun due to the out-flowing solar wind.

Apollo and Amor asteroids are Earth-crossing asteroids that possibly began their lives as comets. Through eons, their coating of ices and gases has been eroded by the sun, exposing what appears to be large chunks of rock (Fig. 167). They are not confined to the asteroid belt as are the great majority of known asteroids, but instead approach or even cross the orbit of the Earth. Usually, within a few tens of millions of years, the Apollos either collide with one of the inner planets or are flung out in wide orbits after a near miss.

Dozens of Apollo asteroids have been identified out of a possible total of perhaps 1,000. Most are quite small and discovered only when they swing close by the Earth (Table 16). Many of these Earth-crossing asteroids do not

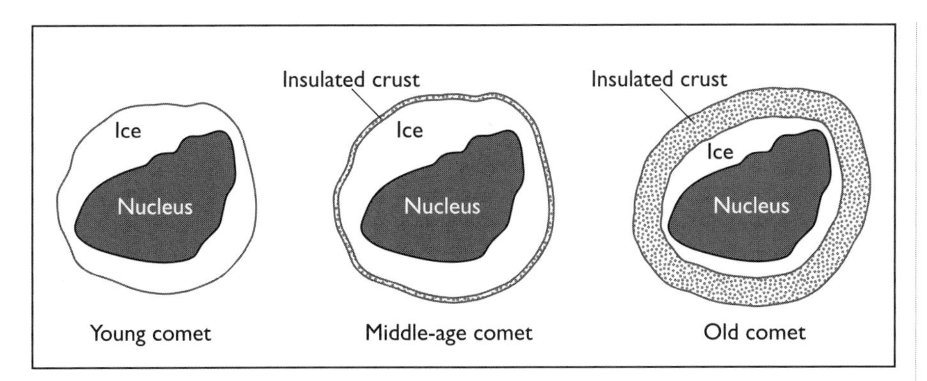

Figure 167 *The life cycle of a comet: When it is young, fresh ice dominates the surface. At middle age, the comet develops an insulated crust. During old age, the crust becomes thick enough to cut off all cometary activity.*

originate in the asteroid belt, but are believed to be comets that have exhausted their volatile material after repeated encounters with the sun and lost their ability to produce a coma or tail. Inevitable collisions with the Earth and the other inner planets steadily depletes them, requiring an ongoing source of new Apollo-type asteroids either from the asteroid belt or from the contribution of burned-out comets.

For a comet to evolve into an asteroid, it must enter a stable orbit in the inner part of the solar system. Meanwhile, its activity is reduced so that it

TABLE 16 CLOSEST CALLS WITH EARTH

Body	Distance in Angstrom Units (Earth-Moon)	Date
1989 FC	0.0046 (1.8)	Mar. 22, 1989
Hermes	0.005 (1.9)	Oct. 30, 1937
Hathor	0.008 (3.1)	Oct. 21, 1976
1988 TA	0.009 (3.5)	Sep. 29, 1988
Comet 1491 II	0.009 (3.5)	Feb. 20, 1491
Lexell	0.015 (5.8)	Jul. 1, 1770
Adonis	0.015 (5.8)	Feb. 7, 1936
1982 DB	0.028 (10.8)	Jan. 23, 1982
1986 JK	0.028 (10.8)	May 28, 1986
Araki-Alcock	0.031 (12.1)	May 11, 1983
Dionysius	0.031 (12.1)	Jun. 19, 1984
Orpheus	0.032 (12.4)	Apr. 13, 1982
Aristaeus	0.032 (12.4)	Apr. 1, 1977
Halley	0.033 (12.8)	Apr. 10, 1837

becomes a burned-out hulk composed mostly of rock. Even when a comet encounters one of the large outer planets such as Jupiter and is trapped in a short-period orbit, its path around the sun is rarely stable. Soon it reencounters the giant planet and is flung back out into deep space, possibly escaping the solar system altogether.

After a comet achieves a stable short-period orbit, it makes repeated passages close to the sun. Every time it passes by the sun, it loses a few feet of its outer layers. The solar wind forces gas and dust particles to stream away from the comet, but the heavier silicate particles are pulled back into the nucleus by the comet's weak gravity. Gradually, an insulating crust forms to protect the icy inner regions of the nucleus from the sun's heat, and the comet ceases its outgassing and masquerades as an asteroid.

CRATERING RATES

One purpose of planetary science is to compare the geologic histories of the planets and their moons by establishing a relative time scale based on the record of impact cratering. Generally, the older the surface, the more craters that are on it. The heavily-cratered lunar highland is the most ancient region on the Moon. It contains a record of intense bombardment around 4 billion years ago. Since then, the number of impacts rapidly declined, and the impact rate has remained relatively low. If the impacts had continued at a high rate throughout Earth's history, the evolution of life would have been substantially altered.

The rate of cratering also appears to differ from one part of the solar system to another. The cratering rates by asteroids and comets along with the total number of craters suggests the average rates over the past few billion years were similar for the Earth, its moon, and the rest of the inner planets. However, the cratering rates for the moons of the outer planets might have been substantially lower than those for the inner solar system. Nevertheless, the size of the craters in the outer solar system is comparable to that of the inner solar system (Fig. 168).

The cratering rates for the Moon and Mars were nearly the same, except that on Mars erosional agents such as wind and ice have erased many of its craters (Fig. 169). Indeed, what appears to be ancient stream channels flowing away from craters indicate that at some time Mars had running water. On the Moon, however, the dominant mechanism for destroying craters is other impacts. The Moon has so much crater overlap, placing the craters in their proper geologic order is often difficult. The impact rates for Mars might have actually been higher than those for the moon, possibly because Mars is much closer to the asteroid belt. Major obliteration events have occurred on Mars as

Figure 168 *The heavily cratered terrain on Mercury from* Mariner 10 *in March 1974.*

(Photo courtesy NASA)

recently as 200 million to 450 million years ago, whereas most of the scarred lunar terrain was produced billions of years ago.

Impact craters on Earth range in age from a few thousand to nearly 2 billion years old. For the past 3 billion years, the cratering rate for the Earth has been fairly constant, with a major impact, resulting in a crater 30 miles or more in diameter, occurring every 50 to 100 million years. As many as three

Figure 169 *A heavily cratered region on Mars showing the effects of wind erosion, from* Viking Orbiter 1 *in June 1980.*

(Photo courtesy NASA)

large meteorite impacts, producing craters with diameters of at least 10 miles, are expected every million years. Major meteorite impacts also appear to be somewhat periodic, occurring every 26 to 32 million years, possibly accounting for the extinction of species on a similar time scale.

More than 150 known impact craters are scattered around the world, most of which are younger than 200 million years (Table 17). Even though the cratering rate has been rather constant during the past 3 billion years, older craters are less abundant because they were destroyed by erosion or sedimentation. Thus far, only about 10 percent of the expected large craters younger

TABLE 17 LOCATION OF MAJOR METEORITE CRATERS OR IMPACT STRUCTURES

Name	Location	Diameter (Feet)
Al Umchaimin	Iraq	10,500
Amak	Aleutian Islands	200
Amguid	Sahara Desert	
Aouelloul	Western Sahara Desert	825
Bagdad	Iraq	650
Boxhole	Central Australia	500
Brent	Ontario, Canada	12,000
Campo del Cielo	Argentina	200
Chubb	Ungava, Canada	11,000
Crooked Creek	Missouri, USA	
Dalgaranga	Western Australia	250
Deep Bay	Saskatchewan, Canada	45,000
Dzioua	Sahara Desert	
Duckwater	Nevada, USA	250
Flynn Creek	Tennessee, USA	10,000
Gulf of St. Lawrence	Canada	
Hagensfjord	Greenland	
Haviland	Kansas, USA	60
Henbury	Central Australia	650
Holleford	Ontario, Canada	8,000
Kaalijarv	Estonia, USSR	300

(continues)

TABLE 17 CONTINUED

Name	Location	Diameter (Feet)
Kentland Dome	Indiana, USA	3,000
Kofels	Austria	13,000
Lake Bosumtwi	Ghana	33,000
Manicouagan Reservoir	Quebec, Canada	200,000
Merewether	Labrador, Canada	500
Meteor Crater	Arizona, USA	4,000
Montagne Noire	France	
Mount Doreen	Central Australia	2,000
Murgab	Tadjikstan, USSR	250
New Quebec	Quebec, Canada	11,000
Nordlinger Ries	Germany	82,500
Odessa	Texas, USA	500
Pretoria Saltpan	South Africa	3,000
Serpent Mound	Ohio, USA	21,000
Sierra Madera	Texas, USA	6,500
Sikhote-Alin	Siberia, USSR	100
Steinheim	Germany	8,250
Talemzane	Algeria	6,000
Tenoumer	Western Sahara Desert	6,000
Vredefort	South Africa	130,000
Wells Creek	Tennessee, USA	16,000
Wolf Creek	Western Australia	3,000

than 100 million years have been discovered. About two-thirds of the known impact craters are located in stable regions known as cratons, composed of strong rocks in the interiors of continents. The cratons experience low rates of erosion and other destructive processes, allowing craters to be preserved for long periods.

METEORITE IMPACTS

The most accepted theory for the origin of meteorites is that they come from a jumble of asteroids in a wide belt lying between the orbits of Mars and Jupiter.

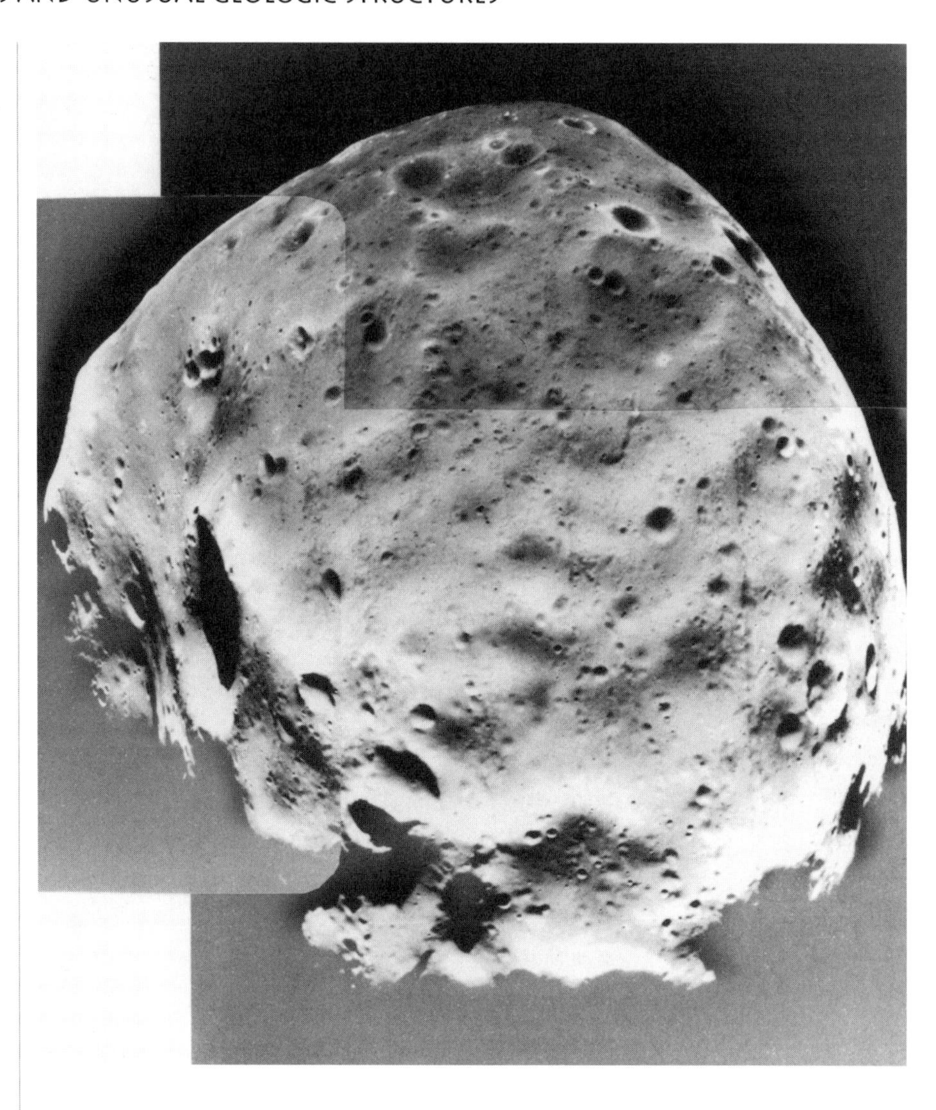

Figure 170 *Mars's moon Phobos, measuring 13 miles across, is thought to be a captured asteroid.*
(Photo courtesy NASA)

Some asteroids might be the rocky cores of dead comets that lost their coating of ice by evaporation and settled into orbit around the sun. A few rare meteorites found on the ice sheets of Antarctica might be pieces of the Martian crust blasted out by large asteroid impacts. Even pieces of the moon might have landed on the Earth when major asteroid impacts blasted them into space.

Asteroids range in size from about a mile to several hundred miles in diameter (Fig. 170). These large chunks along with numerous smaller fragments in the asteroid belt are believed to have a total mass of less than 1 percent of the Earth's mass. Even a 1-mile-wide asteroid could do significant damage upon striking the Earth's surface. However, a mystery remains how these large rock fragments managed to fall into orbits that cross our planet's path. The asteroids seem to be fairly stable and run in nearly circular orbits for millions of years. Then for unknown reasons, their orbits stretch out and become so elliptical that some actually come close to the Earth.

If a large asteroid slammed into the planet, it would expel a huge amount of sediment and produce a deep crater (Fig. 171). The finer material would be lofted high into the atmosphere, while the coarse debris would fall back around the perimeter of the crater, forming a high, steep-banked rim. Not only would rocks be shattered near the impact, but the shock wave passing through the ground would produce shock metamorphism in the surrounding rocks, changing their composition and crystal structure.

The most easily recognizable shock effect is the fracturing of rocks into conical and striated patterns, producing what is called shatter cones due to their conical appearance. They form most readily in fine-grained rocks that have little internal structure, such as limestone and quartzite. Large meteorite impacts also produce shocked quartz grains with prominent striations across crystal faces (Fig. 172). Minerals such as quartz and feldspar develop these features when high-pressure shock waves exert shearing forces on their crystals, producing parallel fracture planes called lamellae.

The high temperatures developed by the force of the impact also fuse sediment into small glassy spherules, which are tiny spherical bodies. Extensive deposits of 3.5-billion-year-old spherules in South Africa are more than a foot thick in places. Spherules of a similar age also have been found in Western Australia. Spherule layers up to 3 feet thick have been found in the Gulf of Mexico and are related to the 65-million-year-old Chicxulub impact structure off Yucatán, Mexico. The spherules resemble the glassy chondrules (rounded granules) in carbonaceous chondrites, which are carbon-rich meteorites, and in lunar soils. The discoveries suggest that massive meteorite bombardments during the early part of the Earth's history played a major role in

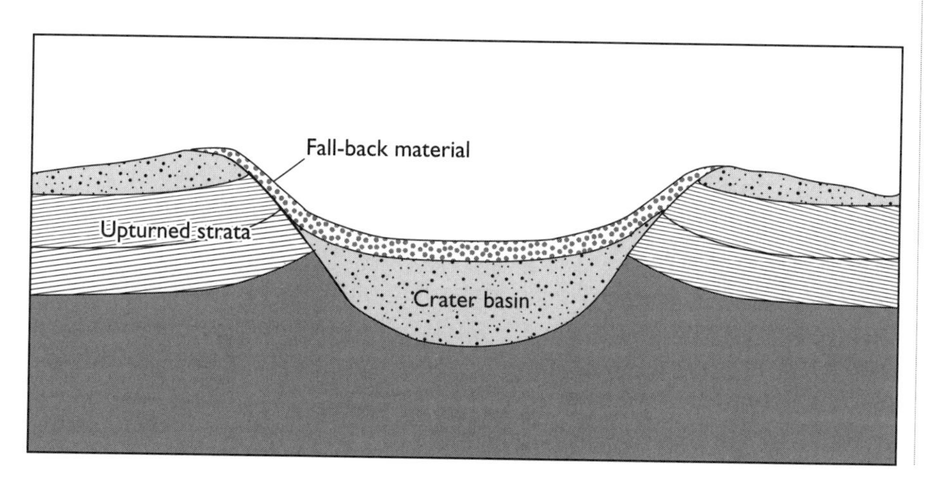

Figure 171 *The structure of a large meteorite crater.*

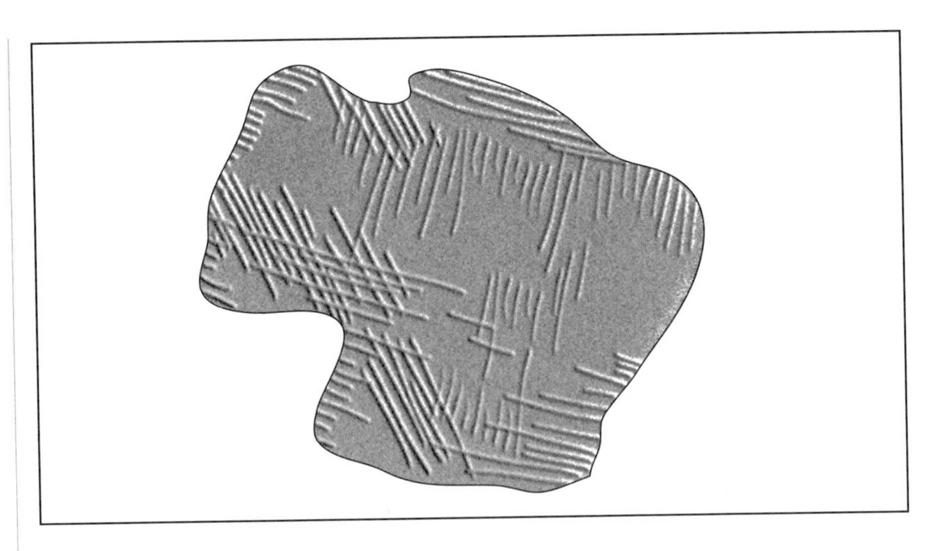

Figure 172 *Lamalle across crystal faces produced by high-pressure shock waves from a large meteorite impact.*

shaping the surface of the planet. In addition, some carbon-rich meteorites might have provided the necessary ingredients for the initiation of life.

Sediments dating 65 million years old located at the boundary between the Cretaceous and Tertiary periods throughout the world (Fig. 173) mark the extinction of the dinosaurs and many other species. They contain shocked quartz grains with distinctive lamellae, common soot from global forest fires set ablaze by glowing bits of impact debris flying past and back through the atmosphere, and unique concentrations of iridium, an isotope of platinum relatively abundant on meteorites and comets but practically nonexistent in the Earth's crust. Two

Figure 173 *The Cretaceous-Tertiary contact is at the base of the white sandstone in the center of the picture, near the foot of the hill, Jefferson County, Colorado.*

(Photo by R. W. Brown, courtesy USGS)

rare amino acids known to exist only on meteorites were also found in the sediment layer. In addition, the sediments contained the mineral stishovite, a dense form of silica found nowhere else on Earth except at known impact sites.

The search for the meteorite impact site has been concentrated around the Caribbean area, where thick deposits of wave-deposited rubble have been found along with melted and crushed rock ejected from the crater. A large asteroid might have struck near the present town of Chicxulub, on the Yucatán Peninsula, Mexico, creating the explosive force of 100 trillion tons of TNT, or 1,000 times more powerful than the detonation of all the world's nuclear arsenals. If the meteorite landed on the seabed just offshore, 65 million years of sedimentation would have long since buried it under thick deposits of sand and mud. Furthermore, a splashdown in the ocean would have created an enormous sea wave, called a tsunami, that would scour the seafloor and deposit its rubble on nearby shores.

CRATER FORMATION

Meteor Crater (Fig. 174), also known as Barringer Crater, 15 miles west of Winslow, Arizona, is one of the most spectacular impact craters on Earth. It is about 4,000 feet across and 560 feet deep and was first mistaken for a volcanic crater because volcanoes were once prevalent in the region. However, its dis-

Figure 174 *Meteor Crater, Coconino County, Arizona.*

(Photo courtesy USGS)

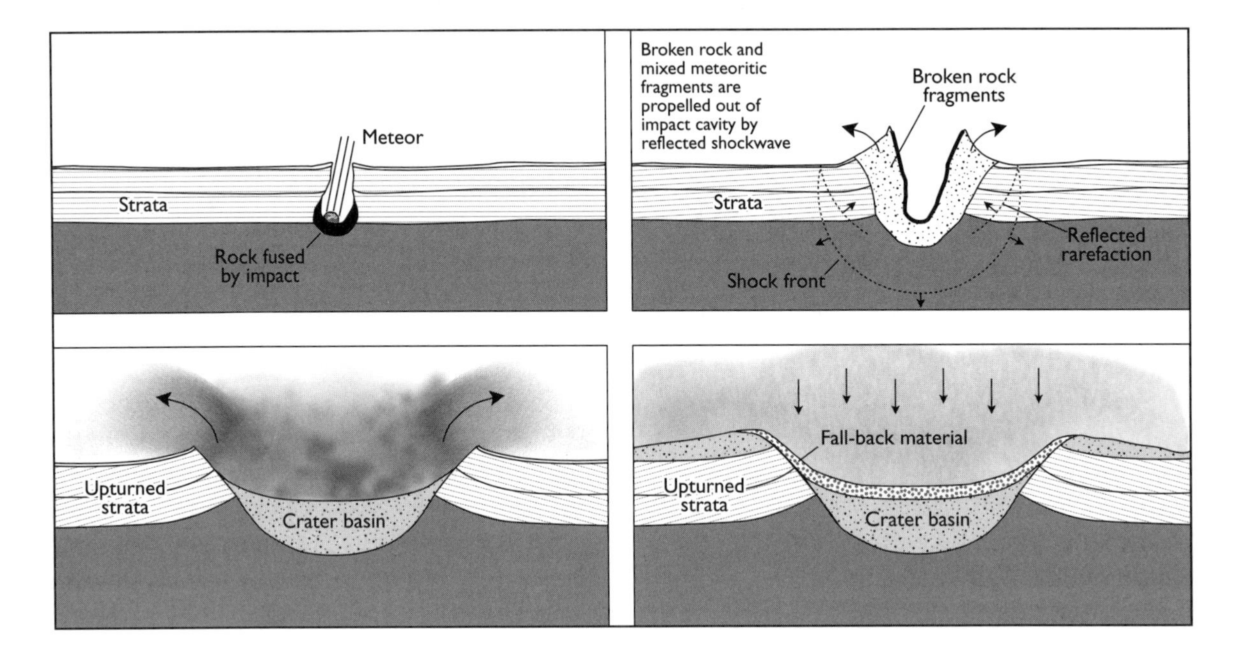

Figure 175 *The formation of a large meteorite crater.*

tinctive appearance more closely resembles craters on the moon. Bore holes drilled in the center of the crater and on the south rim, rising 135 feet above the desert floor, failed to find the meteorite. However, scattered outward from the crater were several tons of metallic meteoritic debris, indicating the meteorite was the iron-nickel variety, measuring about 200 feet in diameter and weighing about 1 million tons. The impactor released the equivalent energy of about 20 megatons of TNT, equal to the most powerful nuclear weapons.

Large meteorites traveling at high velocities completely disintegrate upon impact. In the process, they create craters generally 20 times wider than the meteorites themselves. A large meteorite impact sends out a shock wave with pressures of millions of atmospheres down into the rock and back up into the meteorite. As the meteorite burrows into the ground, it forces the rock aside and flattens itself in the process. It is then deflected and its shattered remains are thrown out of the crater, along with a spray of shock-melted meteorite and melted and vaporized rock that shoots out at a high velocity, leaving behind a deep crater (Fig. 175).

As the spray continues to rise, it forms a rapidly expanding plume. The plume grows to several thousand feet across at the base, while the top extends several miles high into the atmosphere. Most of the surrounding atmosphere is blown away by the tremendous shock wave created by the meteorite impact. The giant plume transforms into an enormous black dust cloud that punches through the atmosphere like the mushroom cloud from a hydrogen bomb explosion. In fact, striking similarities exist between the effects of nuclear detonations and large meteorite impacts.

The crater diameter varies with the type of rock that is impacted due to the relative differences in rock strength. A crater made in crystalline rock can be twice as large as one made in sedimentary rock. Simple craters such as Meteor Crater form deep basins and range up to 2.5 miles in diameter. Larger craters, called complex craters, are much shallower and up to 100 times wider than they are deep. They generally have an uplifted structure in the center surrounded by an annular trough and a fractured rim similar to the central peaks inside craters on the moon as well as those discovered on the seafloor.

IMPACT STRUCTURES

Many impact structures are scattered around the world (Fig. 176). They are large circular features created by the sudden shock of a massive meteorite landing on the surface and are between 1 and 50 miles or more wide. Some impacts form distinctive craters, while others might show only subtle outlines of former craters. In this case, the only evidence of their existence might be a circular disturbed area, where the rocks were altered by shock metamorphism. Shock metamorphism requires the instantaneous application of high temperatures and pressures similar to those found deep in the Earth's interior.

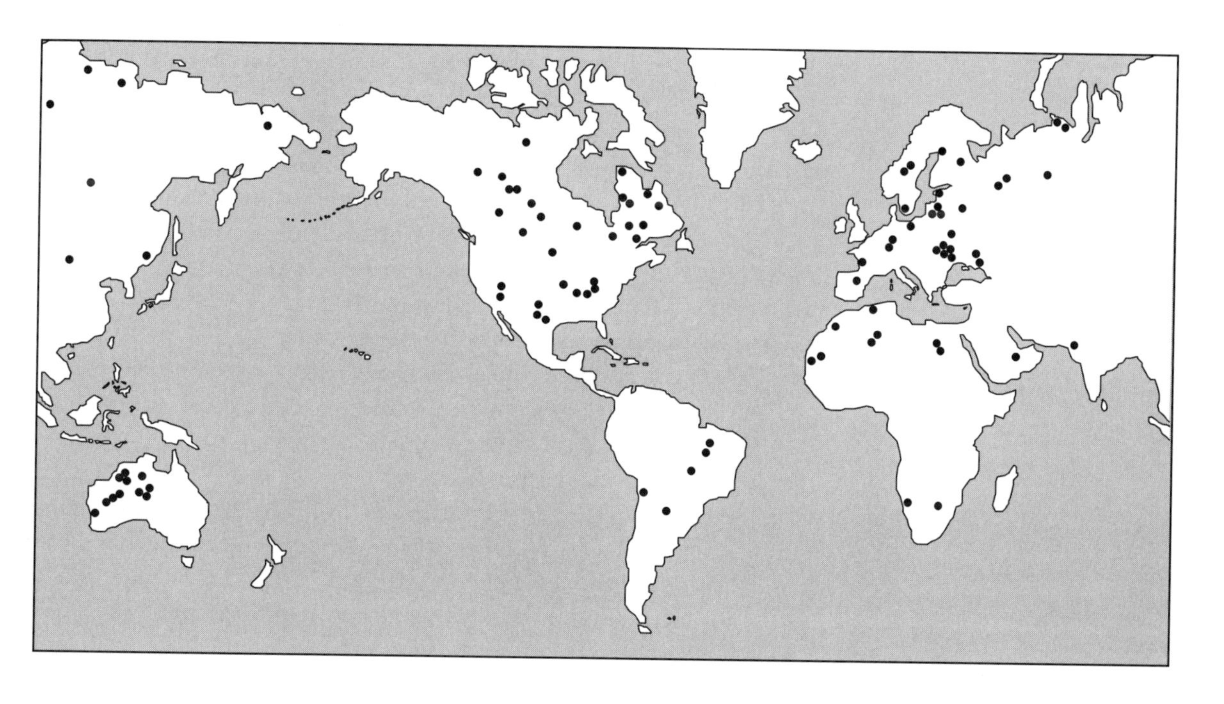

Figure 176 *The locations of some known meteorite craters around the world.*

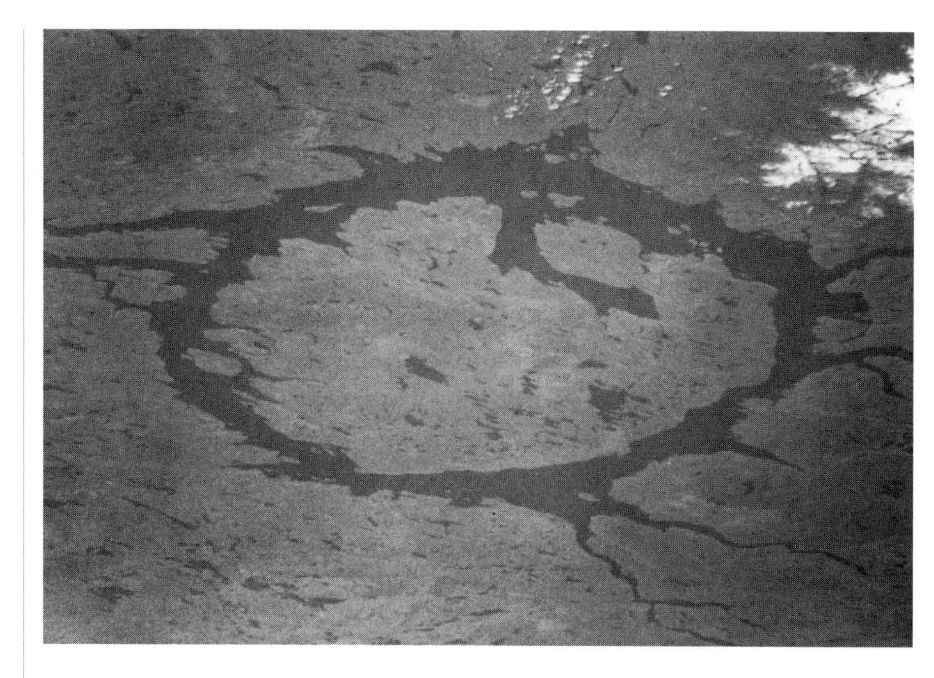

Figure 177 *The Manicouagan impact structure, Quebec, Canada, from* Skylab *in 1973.*
(Photo courtesy NASA)

In Quebec, Canada, the Manicouagan River and its tributaries (Fig. 177) form a reservoir around a roughly circular structure some 60 miles across, making it one of the top six largest craters known on Earth. An almost perfectly circular ring of water, produced when sections of the crater were flooded, surrounds the raised center of the impact structure. The structure is composed of Precambrian rocks reworked by shock metamorphism generated by the impact of a large celestial body.

The 25-mile-wide Saint Martin impact structure northwest of Winnipeg, Manitoba, is mostly hidden beneath younger rocks. Three other impact structures include the 16-mile-wide Rochechouart in France, a 9-mile-wide crater in Ukraine, and a 6-mile-wide crater in North Dakota. All five craters appear to have formed about the same time, roughly 210 million years ago, which coincides with the mass extinctions at the end of the Triassic period that eliminated most reptilian species and paved the way for the dinosaurs.

Many impacts are multiple hits, leaving a chain of two or more craters close together, often produced when an asteroid or comet broke up in outer space or upon entering the atmosphere. A rapid-fire impact from a broken-up object a mile wide apparently produced a string of three 7.5-mile-wide impact craters in the Sahara Desert of northern Chad. Two sets of twin craters, Kara and Ust-Kara in Russia and Gusev and Kamensk near the northern shore of the Black Sea, formed simultaneously only a few tens of miles apart. Splayed across southern Illinois, Missouri, and eastern Kansas are eight large, gently sloping depressions, 2 to 10 miles wide and averaging 60 miles apart. A chain of 10 oblong craters ranging up to 2.5 miles long and a mile wide, running along a 30-mile line near Rio Cuarto, Argentina, suggests a meteorite 500 feet wide hit at a shallow angle and broke into pieces that ricocheted and gouged their way across the landscape roughly 2,000 years ago.

The vast majority of ancient meteorite impacts have long been erased by the Earth's active erosional processes, including the action of rain, wind, glaciers, freezing and thawing, and plant and animal activity. The moon and Mars retain most of their craters because they lack a hydrologic cycle, whose erosional forces wiped out most impact structures on Earth. The forces of erosion have leveled the tallest mountains and gouged out the deepest canyons; no wonder most craters cannot escape these powerful weathering agents. The exceptions are craters in the deserts, which do not receive significant rainfall, or in the Arctic tundra regions, which remain unchanged for ages.

Apparently, very large craters more than 12 miles in diameter and deeper than 2.5 miles are practically impervious to erosion. They escape erosion because the Earth's crust literally floats on a dense, fluid mantle. The process of erosion, whereby material is gradually removed from the continents, is delicately balanced by the forces of buoyancy that keep the crust afloat. Therefore, erosion can only shave off the upper 2 to 3 miles of the continents before the mean height of the crust falls below sea level.

Very large craters are usually deep enough so that even if the entire continent was worn down by erosion, faint remnants would still exist. Craters of extremely large meteorite impacts might temporarily reach depths of 20 miles or more and expose the hot mantle below. The uncovering of the mantle in this manner would result in a gigantic volcanic explosion, releasing tremendous amounts of ash into the atmosphere that would exceed by far all the atmospheric products generated by the meteorite itself.

What appears to be the world's largest impact crater covers most of the western Czech Republic centered near the capital city, Prague. It is about 200 miles in diameter and at least 100 million years old. Concentric circular elevations and depressions surround the city, which is what would be expected if the Prague Basin were indeed a meteorite crater. Moreover, green tektites created by the melt from an impact were found in an arc that follows the southern rim of the basin. The circular outline was discovered in a weather satellite image of Europe and North Africa, and its immense size probably kept it from being noticed earlier.

Several methods can be used for detecting large impact craters that are invisible from the air. Seismic surveys could be used to detect circular distortions in the crust lying beneath thick layers of sediment. The disturbed rock usually produces gravity anomalies that could be detected with gravimeters. The fact that many meteorite falls are of the iron-nickel variety suggests they could be detected by using sensitive magnetic instruments called magnetometers. The surface geology also could indicate areas where rocks were disturbed by the force of the impact or outcrop to form a large circular structure. For example, the 10-mile-wide Wells Creek structure in Tennessee is in an area of essentially flat-lying Paleozoic rocks uplifted to form two concentric synclines (downfolded strata) separated by an anticline (upfolded strata).

METEORITES

Meteorite falls are quite commonplace and have been observed throughout human history. Only meteors of a certain size are sufficiently large to travel all the way through the atmosphere without completely burning up. A meteoroid landing on the Earth's surface is therefore called a meteorite; the suffix *ite* designates it as a rock. Meteorites comprising rock or iron do not appear to originate from the meteoroid streams created by the tails of comets, but instead are fragments of asteroids chipped off by constant collisions.

Historians have often argued that a spectacular meteorite fall of 3,000 stones at l'Aigle in the French region of Normandy in 1803 sparked the early investigation of meteorites. However, this spectacle was actually eclipsed nine years earlier by a massive meteorite shower in Siena, Italy, on June 16, 1794. It was the most significant fall in recent times and spawned the modern science of meteoritics.

The earliest reports of meteorite falls were made by the ancient Chinese during the 17th century B.C. Chinese meteorites are rare, however, and to date no large impact craters have been recognized in China. The oldest meteorite fall, of which material is still being preserved in a museum, is a 120-pound stone that landed outside Ensisheim in Alsace, France, on November 16, 1492. The largest meteorite found in the United States is the 16-ton Willamette Meteorite, which crashed to Earth sometime during the past million years. It was discovered in 1902 near Portland, Oregon, and measured 10 feet long, 7 feet wide, and 4 feet high.

One of the largest meteorites actually seen to fall was an 880-pound stone that landed in a farmer's field near Paragould, Arkansas, on March 27, 1886. The largest known meteorite find, named Hoba West, was located on a farm near Grootfontein, South-West Africa (Namibia), in 1920 and weighs about 60 tons. The heaviest observable stone meteorite landed in a cornfield in Norton County, Kansas, on March 18, 1948. It dug a pit in the ground 3 feet wide and 10 feet deep.

Meteorites are common occurrences, and every day thousands of meteoroids rain on the Earth. Occasionally, meteor showers involve hundreds of thousands of tiny stones. Upward of one million tons of meteoritic material is produced annually. Luckily, most meteors burn up on their journey through the atmosphere, and their ashes contribute to the load of atmospheric dust.

When a meteor explodes near the end of its path through the atmosphere, it produces a bright fireball called a bolide. A magnificent bolide produced the Great Fireball that flashed across the United States on March 24, 1933. Some bolides are bright enough to be seen during the daytime. Occasionally, their explosions can be heard on the ground and might sound like the sonic boom of a supersonic aircraft. Everyday thousands of bolides are estimated to occur around the world, but most go completely unnoticed.

More than 500 major meteorite falls occur each year, most of which plunge into the ocean and accumulate on the seafloor. For most meteorites that land on the surface, the braking action of the atmosphere slows them, so they only bury themselves a short distance. Not all meteorites are hot when they land because the lower atmosphere cools them, and some might even have a layer of frost on their surfaces. Meteorites also can cause a great deal of havoc; many have crashed into houses and automobiles.

The most easily recognizable meteorites are the iron variety, although they represent only about 5 percent of all meteorite falls. They are composed of iron and nickel along with sulfur, carbon, and traces of other elements. Their composition is thought to be similar to that of the Earth's iron core, and indeed they might have once made up the core of a large planetoid that disintegrated long ago. Due to their dense structure, iron meteorites have a good chance of surviving an impact, and most are found by farmers plowing their fields.

One of the best hunting grounds for meteorites happens to be on the glaciers of Antarctica, where the dark stones stand out in stark contrast to the white snow and ice. When meteorites fall on the continent, they are embedded in the moving ice sheets. At places where the glaciers move upward against mountain ranges, the ice evaporates, leaving meteorites exposed on the surface. Some meteorites that have landed on Antarctica are believed to have come from the moon and even as far away as Mars (Fig. 178), when large impacts blasted out chunks of material and hurled them toward the Earth.

Perhaps the world's largest source of meteorites is the Nullarbor Plain, an area of limestone that stretches for 400 miles along the south coast of Western and South Australia. The pale smooth desert plain provides a perfect backdrop for spotting meteorites, which are usually dark brown or black. Since very little erosion takes place, the meteorites are well preserved and are found just where they have landed. More than 1,000 fragments from 150 meteorites that fell during the last 20,000 years have been recovered. One large iron meteorite, the Mundrabilla meteorite, weighs more than 11 tons.

Stony meteorites (Fig. 179) are the most common type and comprise more than 90 percent of all falls. However, because they are similar to Earth materials and therefore erode easily, they are often difficult to find. The meteorites are

Figure 178 *A meteorite recovered from Antarctica in 1981 and thought to be of possible Martian origin.*

(Photo courtesy NASA)

composed of tiny spheres of silicate minerals in a fine-grained matrix. The spheres are known as chondrules, from the Greek *chondros* meaning "grain," and the meteorites themselves are therefore called chondrites. Most chondrites have a chemical composition believed to be similar to rocks in Earth's mantle, which suggests they were once part of a large planetoid that disintegrated eons ago. One of the most important and intriguing varieties of chondrites are the carbonaceous chondrites, which are among the most ancient bodies in the solar system. They also contain carbon compounds that might have provided the precursors of life on Earth.

STREWN FIELDS

Tektites (Fig. 180), from the Greek *tektos* meaning "molten," are glassy bodies, ranging in color from bottle-green to yellow-brown to black. They are usual-

Figure 179 *The Wolf Creek meteorite from Western Australia, showing crack development on cut surface.*

(Photo by G. T. Faust, courtesy USGS)

Figure 180 *A North American tektite found in Texas in November 1985, showing surface erosional and corrosional features.*

(Photo by E. C. T. Chao, courtesy USGS)

ly small, about the size of pebbles, although some have been known to be fist-size. Tektites come in a variety of shapes from irregular to spherical, including ellipsoidal, barrel, pear, dumbbell, or button-shaped. They also possess unique surface markings that appear to have formed while the tektites solidified during their flight through the air.

Tektites have a distinct chemical composition, unlike that of meteorites, that is much like the volcanic glass obsidian but with far less gas and water and no microcrystals, a characteristic that is unknown for any kind of volcanic glass. Tektites contain abundant silica similar to pure quartz sands such as those used in the manufacture of glass. Indeed, tektites appear to be natural glasses formed by the intense heat generated by a large meteorite impact. The impact flings molten material far and wide, and the liquid drops of rock solidify into various shapes while airborne.

Tektites cannot have originated outside the Earth because their composition more closely matches that of terrestrial rocks. Their distribution over the planet's surface suggests they were launched at high velocity by a powerful mechanism, such as a large meteorite impact or a major volcanic eruption. However, terrestrial volcanoes are much too weak to produce the observed strewn fields of tektites that travel nearly halfway around the world (Fig. 181). The largest of these is the Australian strewn field, stretching from the Indian Ocean, southern China, southeastern Asia, Indonesia, the Philippines, and Australia. It comprises perhaps 100 million tons of tektites that are 750,000 years old. The Australian tektites are roundish or chunky objects that show little evidence of internal strain.

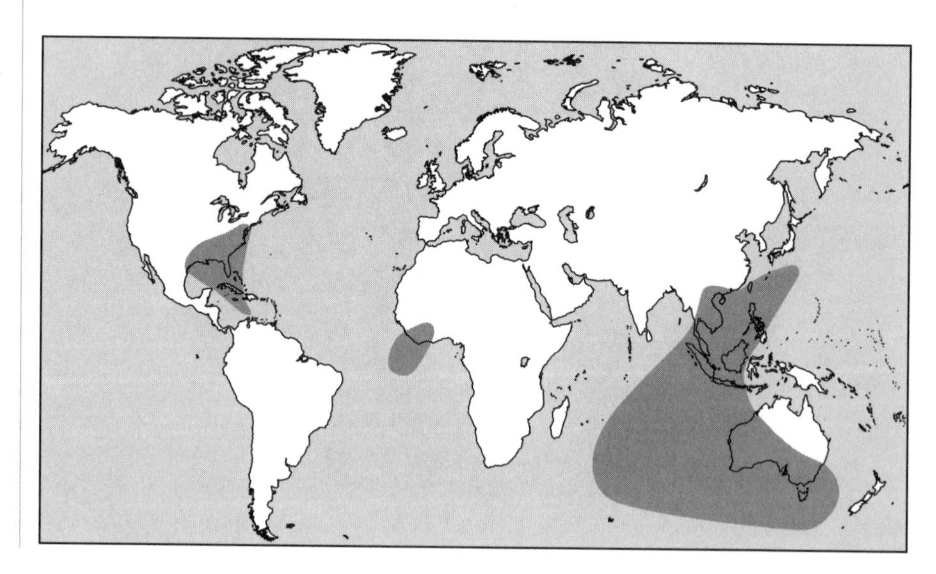

Figure 181 *The distribution of tektites in major strewn fields around the world.*

Mysterious glass fragments strewn over Egypt's Western Desert are believed to be melts from a huge impact that took place about 30 million years ago. Large, fist-size, clear glass fragments are found scattered across the Libyan Desert. Analysis of trace elements indicates the glasses were produced by an impact into the desert sands. The 2-billion-year-old Vredefort structure in South Africa was identified as an impact structure solely from its impact melt and high iridium content.

Shocked metamorphic minerals believed to have originated from a massive impact 65 million years ago, when the dinosaurs became extinct, are strewn across western North America from Canada to Mexico. Shocked quartz and feldspar grains found in the Raton Basin of northeast New Mexico indicates the impact took place either on land or on the continental shelf because these minerals are rarely found in oceanic crust. The large size of the impact grains also suggests the impact occurred nearby, perhaps within 1,000 miles of the strewn field.

In the cornfields of Manson, Iowa, lies a 22-mile-wide crater buried beneath 100 feet of glacial till. This is an ideal site for the giant impact that created the North American strewn field. The rock type at the Manson site is the right composition and age and in the right location to account for the large size of the shocked quartz and feldspar grains in the western United States.

Traces of ancient impact structures also might exist beneath the ocean. Because nearly three-quarters of the Earth is covered by water, most meteorites land in the ocean, and several sites have been selected as possible marine impact craters. The most pronounced undersea impact crater is the 35-mile-wide Montagnais structure 125 miles off the southeast coast of Nova Scotia. The crater is 50 million years old and closely resembles craters on dry land, but its rim is 375 feet beneath the sea and the crater bottom is 9,000 feet deep.

The crater was created by a large meteorite up to 2 miles wide. The impact raised a central peak similar to those seen inside craters on the moon. The structure also contained rocks melted by a sudden shock. Such an impact would have sent a tremendous tidal wave crashing down on nearby shores. The crater is a likely candidate for the source of the North American tektites due to its size and location. However, its age indicates the crater is several million years too young to have been responsible for the tektites. Yet the ocean is vast, and no doubt other craters will be found.

After examining effects of meteorite impacts, the next chapter will look at some of Earth's most unusual structures.

12

UNUSUAL STRUCTURES

CREATION OF UNIQUE ROCK FORMATIONS

This chapter visits rock formations and geologic structures formed by special geologic activities. The Earth's architecture would not be complete without its many unusual structures created by a variety of geologic processes. The surface of the Earth is fashioned by various forces, including the interactions of crustal plates, gravitational impacts, erosion, and collapse. These provide a large variety of unusual geologic structures, ranging from large-scale features in the crust to individual sculptures in stone.

Erosion scours the land as well as the seabed, producing deflation basins and a complex seafloor geology. The expulsion of gases under high pressure produces blowouts both on land and on the seafloor. The Earth hosts a variety of holes in the ground, including potholes, sinkholes, and numerous craters. Among the most unusual structures are fumaroles and geysers, crater lakes, and lava lakes. These are just a few of the many great wonders created by the Earth's active geology.

ROCK MONUMENTS

Out in the rugged American West lie many monuments created when a resistant cap rock protected the sediments below while the surrounding landscape eroded, leaving behind tall pinnacles carved out of stone. Perhaps nowhere is this phenomenon better displayed than at Monument Valley on the border of Arizona and Utah near Four Corners. Instead of crowding together as in other parts of the mile-high Colorado Plateau, however, the spires, mesas, and ragged crags are widely scattered across the desert floor (Fig. 182).

The Four Corners region of Arizona, Utah, Colorado, and New Mexico is characterized by spectacular high desert scenery, including tall mesas, broad valleys, and deep canyons. The most impressive of these is the Grand Canyon of northern Arizona. It is a 250-mile-long, 10-mile-wide, and 1-mile-deep cut in the crust formed by the forces of uplift and erosion as the Colorado River sliced its way through half a billion years of accumulated sediments and

Figure 182 *Monument Valley from the top of Hunts Mesa, Navajo County, Arizona.*

(Photo by I. J. Witkind, courtesy USGS

Precambrian basement rock. Much of the canyon was not carved out by piecemeal erosion grain-by-grain, but by catastrophic landsliding that tore away whole canyon walls (Fig. 183).

The region also comprises great inselbergs, from German meaning "island mountains." These are isolated residual uplands standing separately above the general level of the surrounding plains. Inselbergs are similar to the monadnocks that grace the eastern United States, where lone mountains, hills, and peneplains

Figure 183 *The Grand Canyon, Coconino County, Arizona.*

(Photo courtesy USGS)

Figure 184 *Circle Cliffs upwarp with steeply dipping beds of the Waterpocket Fold, Garfield County, Utah.* (Photo by R. G. Luedke, courtesy USGS)

are the sole highlands. Inselbergs are erosional remnants rising abruptly above the ground and are represented by ridges, domes, or hills. The sharp transition apparently resulted from anomalous weathering patterns related to the rock structure of a particular remnant as well as the topography of the area.

The Four Corners regional structure is predominantly horizontal, with sedimentary strata locally deformed into broad domes, basins, and monoclines, which are steeply inclined sedimentary strata in an area where the bedding is relatively flat-lying. Waterpocket Fold (Fig. 184) near Lake Powell, in southeastern Utah, is among the best examples of a monocline. The flanks of many monoclines present rows of flatirons that stand out from the rest of the formation due to differential weathering of the strata. They resemble old-fashioned steam irons standing side by side. Flatirons are generously displayed in the steeply folded terrain of Utah, Wyoming, and Colorado.

Some monuments stand alone as testaments to the unusual geologic activity that carved them out of the ancient rock. El Capitan Peak in Guadalupe Mountains National Park in west Texas is composed of a massive block of limestone rising high above its sloping flanks. Sometimes a long pinnacle called a chimney rock stands well above its surroundings. One such pinnacle is Chimney Rock National Historic Site near Scotts Bluff, Nebraska, which is named for the 800-foot bluff that stands alone in the middle of the prairie. The town of Chimney Rock in southwest Colorado is named for a tall spire standing nearby.

In northwest New Mexico, a jagged monument called Shiprock rises 1,400 feet above the flat terrain. The volcanic neck is the remnant of an ancient volcano that last erupted more than 30 million years ago. Large dikes radiate outward in three directions like the spokes of a wheel. Similarly, Devils Tower in northeast Wyoming is an eroded volcanic plug rising 1,300 feet above the surrounding plain (Fig. 185). It is composed of solidified magma that once filled the main conduit or feeder pipe of a volcano, and erosion has left the resistant rock standing all alone. Along its flanks, the plug is broken by

Figure 185 *Devils Tower National Monument, Crook County, Wyoming, showing columns and talus.*

(Photo by N. H. Darton, courtesy USGS)

Figure 186 *Columnar jointing in a massive basalt flow at Devils Postpile National Monument, Madera County, California.*

(Photo by F. E. Matthes, courtesy USGS)

columnar jointing created by the shrinking magma as it cooled, creating fractures that shoot through its entire length.

The devil seems to have been at work in California as well. Devils Postpile, southeast of Yosemite National Park, contains rolls upon rolls of six-sided columns in a massive lava flow (Fig. 186). As the lava cooled, it shrank, causing cracking and jointing, which shoot vertically through the entire lava flow. Similar columns stretch across the bare cliffs of the Columbia River basalts, which buried much of Washington, Oregon, and Idaho in basalt floods several thousand feet deep. Early legends have attributed the formation of columnar joints to supernatural forces, as reflected in the names of these sites.

The long, parallel, polygonal columns produced by columnar jointing are seen most frequently in basalts as well as other extrusive and igneous rocks. During the cooling process, a reduction in volume causes a mass of molten rock to develop fractures that divide basaltic lava flows into prismatic columns with hexagonal cross sections similar to honeycombs. The columnar joint patterns evolved from mostly tetragonal shapes to mostly hexagonal ones as the joints grow inward from the flow surfaces as cooling proceeds. The regularity and symmetry of the fracture patterns have long fascinated geologists.

Megalithic monuments such as Stonehenge in southern England and the great statues of Easter Island in the Pacific are among the most dramatic remains of prehistoric culture around the world. Since their discovery, many theories have been put forward to explain their purpose and how they got there, evoking everything from extraterrestrial spaceships to supernatural forces. Many of the hundreds of various circles of tall monoliths found in Europe appear to have been used for astronomical purposes such as the telling of the seasons. The oldest monuments date to about 4,000 B.C. and are often composed of exotic rock hauled in from great distances. Such laborsome activity might be attributed to the worship of stones.

PILLARS OF STONE

Sandstone pillars stretching as much as 190 feet above the desert floor near the town of Gallup in northwestern New Mexico appear to be the world's largest of their type, composed of more than 100 fossilized termite nests. The termite nests are estimated to be 155 million years old and are the first ones found from the Jurassic period. Similar fossilized termite nests have been discovered in Arizona's Petrified Forest National Park dating to the Triassic period 220 million years ago. Interestingly, entomologists, biologists who study insects, have known for some time that termites probably lived early in the Triassic period, just about the time the dinosaurs came onto the scene.

Huge pillars of lava stand like Greek columns on the ocean floor as much as 45 feet tall. How these strange spires formed remains unclear. The best explanation suggests that the pillars were created by the slow advances of lava oozing from volcanic ridges. Several blobs of lava nestle together in a ring, leaving an empty, water-filled space in the center. The sides of these adjoining blobs form the pillar walls, as the outer layers cool on contact with seawater. The insides of the blobs remain fluid until the lava flows back into the vent. The fragile blobs then collapse, somewhat like large empty eggshells, leaving standing behind hollow pillars formed from the interior walls of the ring.

In the petrified forests of Arizona and Wyoming stand tall pillars made from the remains of 200-foot-tall pines that were literally turned to stone. In Yellowstone National Park, the trees stand where they grew, preserved as stony stumps (Fig. 187). A succession of forests have grown on top of each other, forming a layer cake of fossilized stumps 1,200 feet thick. In Petrified Forest of Arizona are found scattered trunks that were carried downstream by ancient floods and buried in the sediment. Groundwater percolating through the sediments replaced wood with silica and erosion has exposed the petrified logs on the surface (Fig. 188).

Bryce Canyon in southern Utah is fashioned out of colorful rocks similar to those that make up the Painted Desert farther to the south in Arizona. Uplift and erosion has carved out a fantastic forest of pinnacles, spires, and columns (Fig. 189) created by a maze of ravines at the edge of a plateau. Similar structures can be found at Devils Half Acre in central Wyoming.

Figure 187 *Petrified tree trunks at Fossil Forest, Specimen Ridge, Yellowstone National Park, Wyoming.*

(Photo by J. P. Iddings, courtesy USGS)

The Canyonlands of Utah and other parts of the West are sprinkled with isolated pinnacles of sandstone called prairie dogs, named so because of the animal's penchant for standing on its haunches to see above the grass on the lookout for predators. Several pinnacles might be clumped together into prairie dog towns. Sometimes erosion has played its usual tricks, and faces and other features can be seen with enough imagination.

Sometimes the statues come complete with their own headgear, including wide sombreros called Mexican hats. Perhaps the most prominent of these

Figure 188 *Petrified tree trunk at the Petrified Forest National Monument, Apache-Navajo Counties, Arizona.*

(Photo by H. E. Gregory, courtesy USGS)

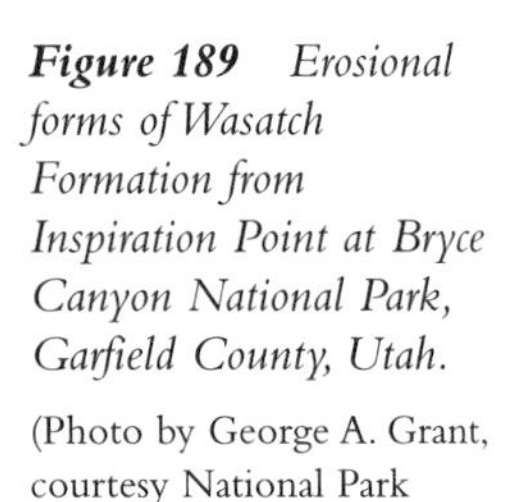

Figure 189 *Erosional forms of Wasatch Formation from Inspiration Point at Bryce Canyon National Park, Garfield County, Utah.*

(Photo by George A. Grant, courtesy National Park Service)

is found in southeast Utah north of Monument Valley near a small town appropriately named Mexican Hat. The hat is formed out of a resistant cap rock that sits precariously on top of an eroded remnant.

A sedimentary dome is created when the crust is heaved upward often due to salt tectonics. Since salt buried in the crust from ancient seabeds is lighter than the surrounding rocks, it slowly rises toward the surface, bulging the overlying strata upward. Often oil and gas are trapped in these structures, and petroleum geologists spend much time looking for salt domes. Upheaval Dome in Canyon Lands National Park, Utah, is perhaps the most compelling example of a huge salt plug that heaved up the overlying strata into a huge bubblelike fold 3 miles wide and 1,500 feet high.

The structure is also interpreted as a deeply eroded astrobleme, which is the remnant of an ancient impact structure gouged out by a large cosmic body striking the Earth such as an asteroid or comet. Erosion has removed as much as a mile or more of the overlying beds since the meteorite smashed into the ground between 30 and 100 million years ago, making the structure possibly the planet's most deeply eroded impact crater.

The original crater apparently made a 4.5-mile-wide hole in the ground that has been heavily modified by deep erosion over many years. The dome itself appears to be a central rebound peak formed when the ground heaved upward by the force of the impact. The meteorite was estimated to be about 1,700 feet wide, and crashed to Earth with a velocity of several thousand miles per hour. On impact, it created a huge fireball that would have incinerated everything for hundreds of miles.

BLOWOUTS

Pockets of gases lie trapped under high pressure deep beneath the floor of the ocean. As the pressure increases, the gases explode undersea, spreading debris far and wide and producing huge craters on the ocean floor. The gases rush to the surface in great masses of bubbles that burst in the open air, resulting in a thick, foamy froth on the surface of the ocean. A ship sailing into such a foamy sea would immediately lose all buoyancy and sink to the bottom since it is no longer supported by seawater.

In 1906, sailors in the Gulf of Mexico actually witnessed a massive gas blowout that sent mounds of bubbles to the surface. At the site, a large crater was discovered on the ocean floor, lying in 7,000 feet of water southeast of the Mississippi River Delta. The elliptical hole measured 1,300 feet long, 900 feet wide, and 200 feet deep and sat atop a small hill. Downslope laid more than 2 million cubic yards of ejected sediment. Apparently, gases seeped upward along cracks in the seafloor and collected under an impermeable barrier. Eventually, the pressure forced the gas to blow off its cover, forming a huge blowout crater.

Beneath the Pacific Ocean near French Polynesia, strange single-frequency notes were found emanating from clouds of bubbles billowing out of undersea volcanoes. The notes were among the purest in the world, far better than those played by any musical instrument. The low frequency of the sound suggested the source had to be quite large. Further search of the ocean depths uncovered a huge swarm of bubbles. When undersea volcanoes gush out magma and scalding water, the surrounding water boils away into bubbles of steam. As the closely packed bubbles rise toward the surface, they rapidly change shape, producing extraordinary single-frequency sound waves.

Another type of blowout is produced by the explosive release of volcanic gases. Hole in the Ground in the Cascade Mountains of Oregon is the site of a gigantic volcanic gas explosion that excavated a huge crater. It is a perfectly circular pit several thousand feet across with a rim raised a few hundred feet above the surrounding terrain. Mysteriously, most of the crater lacks vegetation, possibly due to poisonous gases that spilled out during the eruption. Another similar structure called Ubehebe Crater is one of the most impressive sights in Death Valley. It exploded into existence about 1,000 years ago, when molten basalt contacted the shallow groundwater table and suddenly flashed into steam.

POTHOLES

Potholes are a common sight on existing riverbeds as well as on exposed ancient river bottoms (Fig. 190). They are generally smooth-sided circular or elliptical holes in hard bedrock such as granite or gneiss, which are coarse-grained igneous and metamorphic rocks. Potholes described as "glacial" are often located in the northern regions. The glaciers do not actually cut potholes themselves, but release vast quantities of water when they melt, and river channels overflowing with meltwater cause severe erosion and potholes. Another indirect effect glaciers have on pothole formation occurs during the drainage of large lakes fed by glacial meltwater. The gradients of streams flowing into the lakes greatly steepen, during which they erode downward, cutting a population of potholes.

Potholes have similar shapes, but vary greatly in size, with diameters and depths of up to 5 feet and more. One of the largest potholes is near Archbald, Pennsylvania, and measures 42 feet wide and nearly 50 feet deep. Huge, rounded boulders lying on the bottom were responsible for its creation, as torrents of water from a melting ice age glacier whirled the rocks around in the hole. As the rocks spun round and round, abrasion widened and simultaneously deepened the hole.

Figure 190 *Potholes in granite ledges in the James River, Henrico County, Virginia.*

(Photo by C. K. Wentworth, courtesy USGS)

Another large pothole was formed in the bed of the Deerfield River in Shelburne Falls, Massachusetts, and measures nearly 40 feet across. It is surrounded by several smaller potholes that were cut into hard granite gneiss. Excellent examples of potholes, measuring up to 5 feet in diameter and 30 feet deep, exist on Moss Island in Little Falls, New York. Some of the best potholes are found below dams, which expose the hole-ridden bedrock where rapids once existed.

Most potholes form where water is swiftly flowing and turbulent, such as streams with steep gradients and irregular beds or where a large volume of water is forced through a restricted channel. Such conditions occur when glaciers melt, releasing vast quantities of water that overflow river channels and cause erosion and potholes. Potholes also can form by cavitation caused by the implosion of bubbles in plunge pools below waterfalls. Streambed slopes tend to steepen where water flows across a boundary between hard rock and soft sediment. The stream erodes the sediment more easily than the hard rock, creating a zone of high-speed turbulent rapids and waterfalls.

Fast, turbulent streams carry rocks that abrade the streambed as well as each other, becoming rounded in the process. If irregularities or slight depressions occur on the bedrock's surface, water flowing over them is deflected into turbulent eddy currents or whirlpools. Rocks caught in a whirlpool spin around in one place and rub against the sides and bottom of the depression, as abrasion makes the hole wider and deeper. Often, the round, smooth pebbles and boulders that cut the potholes are found on the bottom of the hole, which is often broader than the top. Some potholes that have formed on overhanging ledges might cut completely through the rock to form short, steeply inclined tunnels.

On the border between Idaho and Montana, a gigantic ice dam held back an enormous body of water called Lake Missoula. Between 15,000 and 13,000 years ago, the dam repeatedly burst, sending massive floods of glacial meltwater gushing toward the Pacific Ocean. Along the way, the floodwaters carved out one of the most peculiar landscapes on Earth, known as the Channeled Scablands, whose distinctive suite of landforms found nowhere else attests to the immense stream power of glacial floods. The Missoula floods scarred the face of the land, carving huge canyons and potholes into thick lava formations. Their presence suggests that catastrophic flooding might account for similar landforms in other parts of the world.

SOD PITS

In the fall of 1984, a large chunk of earth in the shape of a keyhole was discovered in an isolated part of north-central Washington State. The area is known as Haystack Rocks for the house-sized boulders, called irratics, deposited by retreating glaciers at the end of the ice age. The hunk of sod lies upright and measures 10 feet long, 7 feet wide, and 2 feet thick and weighs about 3 tons. It has vertical sides and a flat bottom as though cut out of the ground by a giant cookie cutter. Indeed, the description is so apt the slab has been dubbed "the earth cookie."

The piece of turf lies about 73 feet from a hole in the ground with the same dimensions. Investigators had no doubt the sod came from the hole, which had not been there a month earlier. The grass roots had been ripped out, not cut, as though the plug simply popped out of the ground. A trail of dirt lay in a curved path from the hole to the slab, which was rotated about 20 degrees counterclockwise from the position of the hole from which it came.

No signs of an explosion or any other form of violent activity were found, but the area was hit by a minor earthquake a few days before the plug was discovered. The epicenter was 20 miles away, where the earthquake measured 3.0 magnitude. However, the distant earthquake did not seem to have enough energy to cause this type of disturbance. Just beneath the topsoil at the site lies a hard layer of bedrock that curves slightly downward to form the shape of a shallow bowl. Perhaps this structure could focus the energy of the earthquake in such a way that made the earth jump bodily out of place.

This is not a unique phenomenon, and several similar holes have been discovered in other parts of the world. Earthquakes have been known to toss boulders and even people high into the air. Vertical shock waves from a 1797 earthquake in Ecuador supposedly hurled local citizens 100 feet skyward. In 1978, an earthquake in Utah was blamed for creating a depression 2 feet in diameter by tossing fist-sized clods of dirt as far as 14 feet away. One of the most dramatic examples of this phenomenon occurred during the 1897 Assam earthquake in northeast India. The temblor threw huge clods of earth in every direction, some of which landed with their roots pointing skyward.

Earthquakes have been known to play other tricks. Mysterious mima mounds, which are rounded piles of soil standing up to 10 feet high, are clustered in diverse parts of the world. They have puzzled geologists for more than 150 years and have possibly generated a greater variety of hypotheses than any other geologic feature. Mima mounds form in many earthquake-prone areas with markedly different climates, which suggests they were caused by the vibrating ground during earthquakes. If a thin layer of soil rests on a section of hard rock, certain sections will vibrate more heavily than others where the mounds tend to pile up.

FUMAROLES AND GEYSERS

Alaska's Valley of Ten Thousand Smokes was created when Mount Katmai erupted in June 1912. A series of explosions excavated a depression at the west base of the volcano, whereupon viscous lava rose 800 feet in diameter and 195 feet high. The entire valley became a hardened yellowish orange mass, 12 miles long and 3 miles wide. Thousands of white fumaroles, which are volcanic steam vents responsible for giving the valley its name, gushed out of the ground and shot hot water vapor up to 1,000 feet into the air.

Fumaroles are vents at the Earth's surface that expel hot gases, usually in volcanic regions. They are found on the surface of lava flows, in the calderas and craters of active volcanoes, and in areas where hot intrusive magma bodies such as plutons occur. A special type of fumarole called a spatter cone builds into a mound of lava (Fig. 191). The temperature of the gases within a fumarole can reach 1,000 decrees Celsius. Generally, the bulk of the gases consists of steam and carbon dioxide along with smaller quantities of nitrogen, carbon monoxide, argon, hydrogen, and other gases. In other types of fumaroles, called solfataras, from the Italian word meaning "sulfur mine," sulfur gases predominate.

The term *geyser* comes from the Icelandic word *geysir,* meaning "gusher," which more than adequately describes its behavior. Geysers are often intermittent and explosive, consisting of hot water ejected with great force, generally rising 100 to 200 feet high. The jet of water is usually followed by a column of thundering steam. The record height was set by New Zealand's Waimangu Geyser, which spouted to 1,500 feet in 1904.

Figure 191 *Little Beggar spatter cone on the Volcano House trail near Halemaumau Volcano probably formed in 1874, Hawaii County, Hawaii.*

(Photo by H.T. Sterns, courtesy USGS)

The primary requirement for the production of fumaroles and geysers is for a large, slowly cooling magma body to lie near the surface where it provides a continuous supply of heat. The hot water and steam are derived either from juvenile water released directly from magma melts along with other volatiles, or from groundwater that percolates downward near a magma body, where it is heated by convection currents. Volatiles released from the magma body also can heat the groundwater from below.

The tube leading to the surface from a deep underground geyser chamber is often restricted or crooked like the drain pipe under a sink. When water seeps into the chamber from a water table, it is heated from the bottom up. The overlying weight of the water places great pressure on the water in the bottom of the chamber, keeping it from boiling, even though temperatures might greatly exceed the normal boiling point of water. As the water temperature gradually increases, some water near the top of the geyser tube boils off, decreasing the weight and causing the water in the bottom of the chamber to flash into steam. This overcomes the restriction, and hot water and superheated steam gush out of the vent (Fig. 192).

Beneath the huge Yellowstone caldera created by a massive volcanic eruption 600,000 years ago is a volcanic hot spot. It is responsible for the continuous thermal activity that produces a multitude of geysers such as Old

Faithful, whose nearly hourly eruptions can last five minutes and spout a column of steam 130 feet high. In addition, a variety of boiling mud pits and hot water streams are produced when rainwater seeps into the ground, acquires heat from a magma chamber, and rises through fissures in the torn crust. The region is frequently shaken by earthquakes, the largest of which occurred in 1959 and threw off Old Faithful's dependable timing.

In rapidly spreading rift systems such as the East Pacific Rise south of Baja California are hydrothermal vents that built forests of tall chimneys often with branching pipes. Exquisite chimneys up to 30 feet tall called black smokers spew hot water blackened with sulfide minerals into the near-freezing deep abyss. Seawater seeping through the ocean crust is heated near magma chambers below the rifts and is expelled with considerable force through vents like undersea geysers (Fig. 193).

The openings of the vents typically range from less than a half inch to more than 6 feet across. They are common throughout the world's oceans along the midocean spreading ridge system and are believed to be the main route through which the Earth's interior loses heat. The vents exhibit a strange phenomena by glowing in the pitch-black dark, possibly caused by the sudden cooling of the 350-degree water, which produces crystalloluminescence as dissolved minerals crystallize and drop out of solution, thereby emitting light. The light, although extremely dim, is apparently bright enough to allow photosynthesis to take place even on the very bottom of the deep sea.

Figure 192 *An explosive burst of a geyser at Yellowstone National Park, Wyoming.*

(Photo by D. E. White, courtesy USGS)

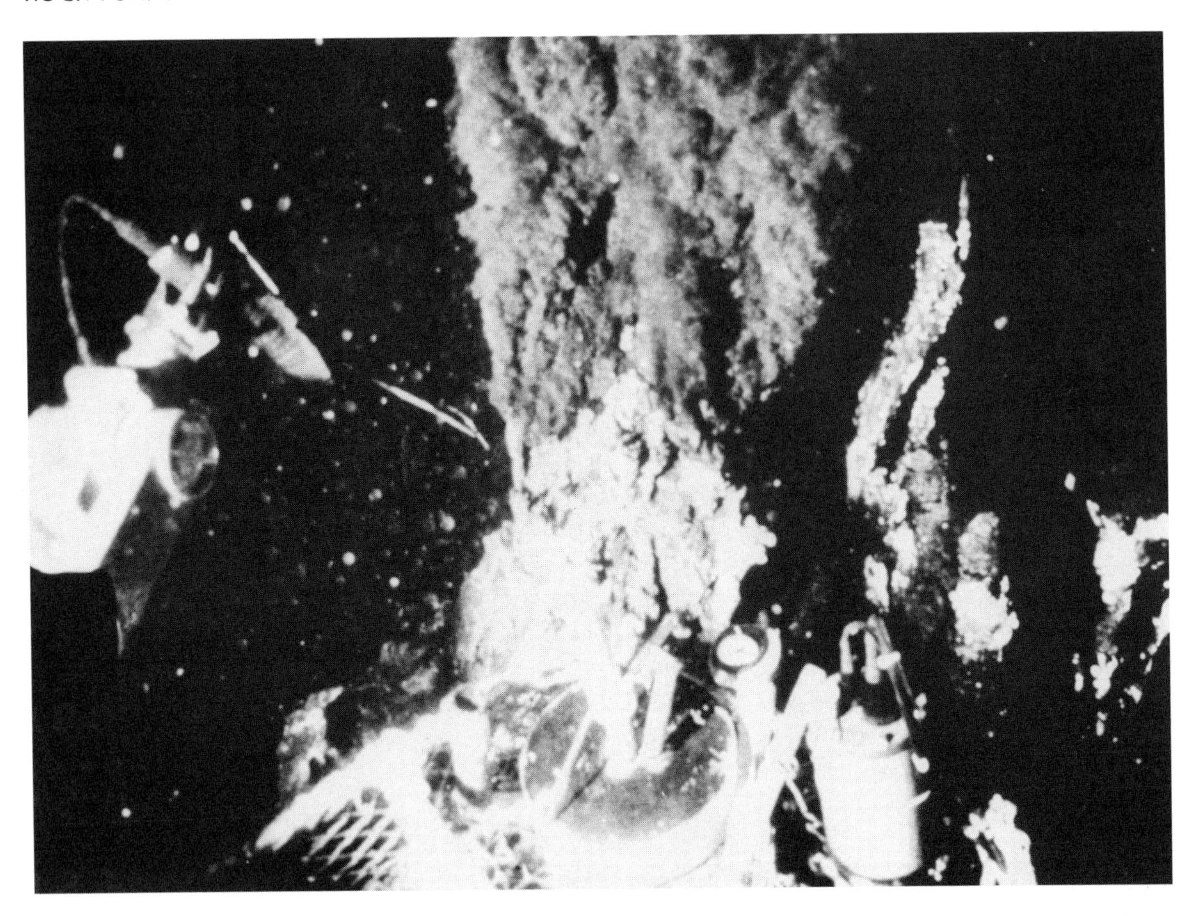

Figure 193 *A hydrothermal vent with sulfide-laden hot water pouring out into cold seawater on the ocean floor. The photograph is taken from the deep-sea submersible* Alvin, *whose claw holds a temperature probe.*

(Photo by N. P. Edgar, courtesy of USGS)

The ghostly white chimneys of the Mariana seamounts in the western Pacific near the world's deepest trench are composed of a form of aragonite, a white calcium carbonate rock with a very unusual texture that normally dissolves in seawater at these great depths. Hundreds of corroded and dead carbonate chimneys were strewn across the ocean floor in wide "graveyards." The fluid temperatures in subduction zones are cool compared with those associated with midocean ridges, enabling crystals of aragonite and calcite to form. Apparently, cool water slowly seeping from beneath the surface enables the carbonate chimneys to grow and avoid dissolution by seawater. Many carbonate chimneys are thin and generally less than 6 feet high. Other chimney structures are thicker and much taller, and occasionally cluster together into ramparts.

Tufa is a porous rock composed of calcite or silica that commonly occurs as an incrustation around the mouths of hot springs. However, in southwestern Greenland, more than 500 giant towers of tufa cluster together in the chilly waters of Ikka Fjord. Some reach as high as 60 feet, and their tops are visible at low tide. The towers are made of an unusual form of calcium carbonate called ikaite, whose crystals form when carbonate-rich water from springs beneath the fjord seeps upward and comes into contact with cold, calcium-laden seawater. Because of the low temperature, the water cannot escape during the precipitation of the mineral and is incorporated into the crystal lattice, producing weird, yet beautiful formations.

About 50 miles west of the Mariana Trench, the world's deepest depression, lies a cluster of large seamounts 2.5 miles below the surface of the sea in a zone about 600 miles long and 60 miles wide. The undersea mountains were built not by hot volcanic rock like most Pacific seamounts, but by cold serpentine, which is a soft, mottled green rock similar to the color of a serpent, hence its name. Serpentine is a low-grade metamorphic rock and the main mineral in asbestos. It originates from the reaction of water with olivine, an olive-green, iron and magnesium-rich silicate and a major constituent of the upper mantle. The erupting serpentine rock flows down the flanks of the seamounts like lava from a volcano and forms gently sloping structures. Many rise over a mile above the ocean floor and measure as much as 20 miles across at the base.

Mud volcanoes exist in many places around the world and usually develop above rising blobs of salt or near ocean trenches. However, beneath the chilly Arctic waters lies a strange mud volcano that spews out a slurry of seafloor sediments mixed with water. It is a half-mile-wide circular feature that lies 4,000 feet deep and is draped by an unusual layer of snowlike natural gas called methane hydrate. The underwater volcanic structure is the first of its kind found covered with such an icy coating draped across a warm mud volcano. Methane hydrate is a solid mass formed when high pressures and low temperatures squeeze water molecules into a crystalline cage around a methane molecule. Vast deposits of methane hydrate are thought to be buried in the ocean floor around the continents and represent the largest untapped source of fossil fuel left on Earth.

CRATER LAKES

When a dormant caldera fills with fresh water from melting snow or rain, it forms a crater lake, whose deepness is determined by the depth of the caldera floor and the water level below the rim. Sediment eroded from the wall of the caldera accumulates into thick deposits on the bottom of the lake. The erosion also widens the caldera. Sometimes resurgence of the caldera floor will create an island in the middle of the lake, capped with young lake sediments.

Crater Lake in Oregon (Fig. 194) originated when the upper 5,000 feet of the 12,000-foot composite cone of Mount Mazama collapsed 6,000 years ago and filled with rainwater and melting snow. It is 6 miles wide and 2,000 feet deep, the sixth deepest lake in the world. The rim of the caldera rises 500 to 750 feet

Figure 194 *Crater Lake in Klammath County, Oregon, formed by the collapse of Mount Mazama 6,000 years ago.*
(Courtesy USGS)

above the surface of the lake. At one end is a small volcanic peak called Wizard Island, which evolved from later volcanic activity. A similar crater lake was formed when the top 1,200 feet of Alaska's Mount Katmai was blown away by an explosive eruption in June 1912. The eruption created a caldera 1.5 miles wide and 2,000 feet deep filled with water from melting snow.

The world's largest crater lake is in northern Sumatra and fills the Toba Caldera, created by the greatest volcanic eruption in 2 million years. The caldera, which extends for nearly 60 miles in its longest dimension, resulted when the roof collapsed over a large magma chamber 75,000 years ago. The floor of the caldera consequently subsided more than a mile, allowing a deep lake to form. Later, the floor heaved upward several hundred feet like a huge piston. The resurgence of the lake bottom formed a 250-square-mile island in the middle of the lake, called Samosir, which might still be rising.

Lake Nios is a crater lake in Cameroon, Africa, that exploded on August 21, 1986, sending a wall of toxic fumes cascading down the hillside, killing 1,700 people along with many animals. The disaster might have been triggered by a small earth tremor that cracked open the deep lake bottom, releasing volcanic gases under high pressure. This created a huge bubble that burst explosively through the surface of the water, churning the clear blue lake to a murky reddish brown from stirred-up bottom sediments. The heavy gases swept down the hillside and spread out in a low-hanging blanket for more than 3 miles downwind from the lake, asphyxiating its victims with a brew of deadly gases.

LAVA LAKES

Perhaps nowhere is the pulse of the planet stronger than in the erupting Kilauea Volcano on Hawaii. Every day, several hundred thousand cubic yards of molten rock gush from a rift zone along its flanks. When the lava has run its course down the mountainside, it flows into the ocean, adding acres of new land to the island. The source of these fiery conditions is a mantle plume of hot rock burning through the Pacific plate from deep inside the Earth. The hot rock has fueled the five volcanoes that build the big Island of Hawaii.

The oldest volcano, Kohala, on the northernmost part of the island last erupted about 60,000 years ago, and today it is worn and eroded, with its northeastern flank deeply incised by spectacular valleys and gorges. Just to the south stands Mauna Kea, which happens to be the tallest mountain on Earth, rising more than 6 miles from the ocean floor. Southwest of Mauna Kea lies Hualalai Volcano, which last erupted in 1801 and is still poised for another outburst. Southeast of Hualalai is Mauna Loa, the world's largest shield volcano. It consists of some 24,000 cubic miles of lava that built up flow upon flow into a huge gently sloping mound, making it the most voluminous mountain on Earth. The youngest volcano, Kilauea, emerges from the side of Mauna Loa. Lava has erupted continuously since the early 1980s from a rift zone on Kilauea, which over time could greatly outgrow its host volcano.

Deep lava lakes on Mount Kilauea are filled with molten basalt at 1,200 decrees Celsius. Basalt is the most common igneous rock formed by the solidification of magma extruded onto the surface of the Earth from great depths. Most of the more than 500 active volcanoes are entirely or predominantly basaltic, including those that built the Hawaiian Islands. At the summit of most volcanoes is a steep-walled depression called a volcanic crater. The crater is connected to the magma chamber by a pipe or vent. When fluid magma moves up the pipe, the lava is stored in the crater until it fills and overflows. During periods of inactivity, back flow can completely drain the crater.

Kilauea rises three-quarters of a mile above sea level and is shaped like an inverted saucer, with a large crater at the summit, from which two rift zones radiate. The eruptions are usually limited to the crater and the rift zones, particularly the eastern rift zone and the fire pit in the crater. On average, Kilauea has erupted at least once a year since 1952. The lava lakes on Kilauea are basalt flows from previous eruptions trapped in large pools and do not solidify to any large extent (Fig. 195).

Figure 195 *Lava pool and fountains on Kilauea, Hawaii in July 1961.*

(Photo by D. H. Richter, courtesy USGS)

Figure 196 *Lava fountain and lake at Kilauea Volcano during 1959 and 1960 eruptions.*

(Photo by D. H. Richter, courtesy USGS)

The depth of the lakes can be substantial, as much as 400 feet for the Kilauea Iki Crater. The lakes take a long time to cool and solidify, generally up to a year for shallow lakes to as long as 25 years for the deepest at Kilauea Iki. Eventually, the natural dikes that channel the lava into the basin collapse, and the lake is cut off from its sources and begins to solidify from the bottom up as well as from the top down. Some lava lakes disappear completely down the bottom of the crater as though the drain plug been pulled.

Drilling into the buried masses of molten rock on Kilauea Iki lava lake was conducted to test its geothermal energy potential. This great lake formed during the 1959 eruption of Kilauea, when lava flows ponded in an ancient crater to depths of 325 feet. Lava happens to be an excellent insulator, and holes drilled 20 years later encountered near molten rock at depths below 160 feet. A single eruption of Kilauea could supply the energy equal to two-fifths the power requirements of the entire United States during the time of the eruption.

The lava lakes often erupt in vigorous fire fountains, which are great volumes of molten basalt sprayed high into the air (Fig. 196). Mauna Loa is known for its tall fountains of white-hot lava that shoot several hundred feet high, forming a characteristic "curtain of fire." Despite their spectacular and violent outbursts, the eruptions are relatively harmless and a great delight to tourists. Their great leaps of fire and smell of brimstone is why such volcanoes are often referred to as "gateways to Hell."

CONCLUSION

A large part of understanding how geology works is comprehending the development of the Earth's landforms. Erosion and sedimentation are important fundamental processes of geology. As sediments flow into ocean basins, they build up layer by layer like a thick stack of paper. Uplift and stream erosion carves this layer cake of sediments into canyons and valleys. Some canyons might cut so deep, they uncover the basement rocks that formed the Earth's earliest crust.

When molten rock invades the sediment pile, it forms large granitic bodies buried deep below. If erosion peels away the sediments from around these massive rocks, they stand as majestic mountain ranges. If the molten rock makes its way to the surface, it erupts as a volcano. Glaciers sprawling over the mountains gouge out large chunks of rock and meltwater deposits the sediments into heaps of glacial till.

Great tectonic pressures pressing against the veneer of sediments as continents collide, form huge folds and fractures in the crust. When the faults slip, they produce powerful earthquakes that cause landslides and other earth movements, further sculpting the landscape. The fold belts become huge mountains riddled with complex drainage systems. Erosion tears down the mountains and rivers carry the sediments to the sea, and the whole process starts over again.

GLOSSARY

aa lava (AH-ah) a lava that forms large jagged, irregular blocks
abrasion erosion by friction, generally caused by rock particles carried by running water, ice, and wind
abyss (ah-BIS) the deep ocean, generally over a mile in depth
aerosol a mass of minute solid or liquid particles dispersed in the air
agglomerate (ah-GLOM-eh-ret) a pyroclastic rock composed of consolidated volcanic fragments
albedo the amount of sunlight reflected from an object and dependent on its color and texture
alluvium (ah-LUE-vee-um) stream-deposited sediment
alpine glacier a mountain glacier or a glacier in a mountain valley
andesite an intermediate type of volcanic rock between basalt and rhyolite
anticline folded sediments that slope downward away from a central axis
Apollo asteroids asteroids that come from the main belt between Mars and Jupiter and cross the Earth's orbit
aquifer (AH-kwe-fer) a subterranean bed of sediments through which groundwater flows
aragonite (ah-RAG-gone-ite) a calcium carbonate mineral similar to calcite and found in cave and hot springs deposits
arches archlike features in rock formed by erosion
arête (ah-RATE) a sharp-crested ridge formed by abutting cirques
arkose a feldspar-rich sandstone
ash fall the fallout of small, solid particles from a volcanic eruption cloud

asperite (AS-per-ite) the point where a fault hangs up and eventually slips causing earthquakes

asteroid a rocky or metallic body whose impact on the Earth creates a large meteorite crater

asteroid belt a band of asteroids orbiting the sun between the orbits of Mars and Jupiter

asthenosphere (as-THE-nah-sfir) a layer of the upper mantle from about 60 to 200 miles below the surface that is more plastic than the rock above and below and might be in convective motion

astrobleme (as-TRA-bleem) eroded remains on the Earth's surface of an ancient impact structure produced by a large cosmic body

avalanche (AH-vah-launch) a slide on a snowbank triggered by vibrations from earthquakes and storms

back-arc basin a seafloor spreading system of volcanoes caused by an extension behind an island arc that is above a subduction zone

Baltica (BAL-tik-ah) an ancient Paleozoic continent of Europe

barrier island a low, elongated coastal island that parallels the shoreline and protects the beach from storms

basalt (bah-SALT) a dark volcanic rock that is usually quite fluid in the molten state

basement rock subterranean igneous, metamorphic, granitized, or highly deformed rock underlying younger sediments

batholith (BA-the-lith) the largest of intrusive igneous bodies and more than 40 square miles on its uppermost surface

bedrock solid layers of rock beneath younger materials

black smoker superheated hydrothermal water rising to the surface at a midocean ridge. The water is supersaturated with metals, and when exiting through the seafloor it quickly cools and the dissolved metals precipitate, resulting in black, smokelike effluent.

blowout a hollow caused by wind erosion or a crater on the seafloor formed by expelled gasses

blue hole a water-filled sinkhole

blueschist (BLUE-shist) metamorphosed rocks of subducted ocean crust exposed on land

bolide an exploding meteor whose fireball is often accompanied by a bright light and sound when passing through the Earth's atmosphere

breccia (BRE-cha) a rock composed of angular fragments in a fine-grained matrix

butte a flat-topped hill with steep slopes

calcite a mineral composed of calcium carbonate

caldera (kal-DER-eh) a large pitlike depression at the summits of some volcanoes and formed by great explosive activity and collapse

calving formation of icebergs by glaciers breaking off upon entering the ocean

carbonaceous (kar-bah-NAY-shus) a substance containing carbon, namely sedimentary rocks such as limestone and certain types of meteorites

carbonate a mineral containing calcium carbonate such as limestone

carbon cycle the flow of carbon into the atmosphere and ocean, the conversion to carbonate rock, and the return to the atmosphere by volcanoes

catchment area the recharge area of an groundwater aquifer

Cenozoic (SIN-eh-zoe-ik) an era of geologic time comprising the last 65 million years

chalk a soft form of limestone composed chiefly of calcite shells of microorganisms

chert an extremely hard, fine-grained quartz mineral

chondrite (KON-drite) the most common type of meteorite, composed mostly of rocky material with small spherical grains

chondrule (KON-drule) rounded granules of olivine and pyroxine found in stony meteorites called chondrites

circum-Pacific active seismic regions around the rim of the Pacific plate coinciding with the Ring of Fire

cirque (serk) a glacial erosional feature, producing an amphitheaterlike head of a glacial valley

col (call) a saddle-shaped mountain pass formed by two opposing cirques

coma the atmosphere surrounding a comet when it comes within the inner solar system. The gases and dust particles are blown outward by the solar wind to form the comet's tail

comet a celestial body believed to originate from a cloud of similar bodies that surround the Sun and develop a long tail of gas and dust particles when traveling near the inner solar system

conduit a passageway leading from a reservoir of magma to the surface of the Earth through which volcanic products pass

cone, volcanic the general term applied to any volcanic mountain with a conical shape

conglomerate (KON-glom-er-ate) a sedimentary rock composed of welded fine-grained and coarse-grained rock fragments

continent a landmass composed of light, granitic rock that rides on the denser rocks of the upper mantle

continental drift the concept that the continents have been drifting across the surface of the Earth throughout geologic time

continental glacier an ice sheet covering a portion of a continent

continental margin the area between the shoreline and the abyss that represents the true edge of a continent

continental shelf the offshore area of a continent in shallow sea

continental shield ancient crustal rocks upon which the continents grew

continental slope the transition from the continental shelf to the deep-sea basin

convection a circular, vertical flow of a fluid medium by heating from below. As materials are heated, they become less dense and rise, cool, become denser again, and sink

convergent plate the boundary between crustal plates where the plates come together; generally corresponds to the deep-sea trenches where old crust is destroyed in subduction zones

coquina (koh-KEY-nah) a limestone comprised mostly of broken pieces of marine fossils

coral any of a large group of shallow-water, bottom-dwelling marine invertebrates that are reef-building colonies common in warm waters

Cordillera (kor-dil-ER-ah) a range of mountains that includes the Rockies, Cascades, and Sierra Nevada in North America and the Andes in South America

core the central part of the Earth, consisting of a heavy iron-nickel alloy; also, a cylindrical rock sample drilled through the crust

correlation (KOR-eh-LAY-shen) the tracing of equivalent rock exposures over distance usually with the aid of fossils

crater, meteoritic a depression in the crust produced by the bombardment of a meteorite

crater, volcanic the inverted conical depression found at the summit of most volcanoes, formed by the explosive emission of volcanic ejecta

craton (CRAY-ton) the stable interior of a continent, usually composed of the oldest rocks

creep the slow flowage of earth materials

crevasse (kri-VAS) a deep fissure in the crust of a glacier

crust the outer layers of a planet's or a moon's rocks

crustal plate a segment of the lithosphere involved in the interaction of other plates in tectonic activity

delta a wedge-shaped layer of sediments deposited at the mouth of a river

desertification (di-zer-te-fa-KA-shen) the process of becoming arid land

desiccated basin (de-si-KAY-ted) a basin formed when an ancient sea evaporated

diapir (DIE-ah-per) the buoyant rise of a molten rock through heavier rock

diatom (DIE-ah-tom) any of numerous microscopic unicellular marine or freshwater algae having siliceous cell walls

dike a tabular intrusive body that cuts across older strata

divergent plate the boundary between crustal plates where the plates move apart, it generally corresponds to the midocean ridges where new crust is formed by the solidification of liquid rock rising from below

dolomite (DOE-leh-mite) a sedimentary rock formed by the replacement of calcium with magnesium in limestone

domepit a vertical shaft connecting different levels in a cave

dropstone a boulder embedded in an iceberg and dropped to the seabed upon melting

drumlin a hill of glacial debris facing in the direction of glacial movement

dune a ridge of wind-blown sediments usually in motion

earth flow the downslope movement of soil and rock

earthquake the sudden rupture of rocks along active faults in response to geological forces within the Earth

East Pacific Rise a midocean spreading center that runs north-south along the eastern side of the Pacific and the predominant location upon which the hot springs and black smokers have been discovered

elastic rebound theory the theory that earthquakes depend on rock elasticity

eolian (ee-OH-lee-an) a deposit of wind-blown sediment

epicenter the point on the Earth's surface directly above the focus of an earthquake

erosion the wearing away of surface materials by natural agents such as wind and water

erratic boulder a glacially deposited boulder far from its source

escarpment (es-KARP-ment) a mountain wall produced by the elevation of a block of land

esker (ES-ker) a curved ridge of glacially deposited material

evaporite (ee-VA-per-ite) the deposition of salt, anhydrite, and gypsum from evaporation in an enclosed basin of stranded seawater

exfoliation (eks-foe-lee-A-shen) the weathering of rock causing the outer layers to flake off

extrusive (ik-STRU-siv) an igneous volcanic rock ejected onto the surface of a planet or moon

facies an assemblage of rock units deposited in a certain environment

fault a break in crustal rocks caused by earth movements

feldspar (FELD-spar) the most abundant rocks in the Earth's crust, composed of silicates of calcium, potassium, and sodium

fissure a large crack in the crust through which magma might escape from a volcano

fjord (fee-ORD) a long, narrow, steep-sided inlet of a mountainous, glaciated coast

floodplain the land adjacent to a river that floods during river overflows

flowstone a mineral deposit formed on the walls and floor of a cave

fluvial (FLUE-vee-al) pertaining to being deposited by a river

foraminifer (for-eh-MI-neh-fer) a calcium carbonate secreting organism that lives in the surface waters of the oceans. After death, their shells form the primary constituent of limestone and sediments deposited on the seafloor

formation a combination of rock units that can be traced over distance

fossil any remains, impression, or trace in rock of a plant or animal of a previous geologic age

frost heaving the lifting of rocks to the surface by the expansion of freezing water

frost polygons polygonal patterns of rocks from repeated freezing

fumarole (FUME-ah-role) a vent through which steam or other hot gases escape from underground such as a geyser

gabbro (GA-broe) a dark, coarse-grained intrusive igneous rock

geomorphology (JEE-eh-more-FAH-leh-jee) the study of surface features of the Earth

geosyncline (geo-SIN-kline) a basinlike or elongated subsidence of the Earth's crust. Its length might extend for several thousand miles and might contain sediments thousands of feet thick, representing millions of years of deposits. A geosyncline generally forms along continental edges and is destroyed during periods of crustal deformation

geothermal the generation of hot water or steam by hot rocks in the Earth's interior

geyser (GUY-sir) a spring that ejects intermittent jets of steam and hot water

glacier a thick mass of moving ice occurring where winter snowfall exceeds summer melting

glacier burst a flood caused by an underglacier volcanic eruption

glaciere (GLAY-sher-ee) an underground ice formation

Glossopteris (GLOS-op-ter-is) late Paleozoic fern living on Gondwana

gneiss (nise) a banded, coarse-grained metamorphic rock with alternating layers of different minerals, consisting of essentially the same components as granite

Gondwana (gone-DWAN-ah) a southern supercontinent of Paleozoic time, comprising Africa, South America, India, Australia, and Antarctica, which broke up into the present continents during the Mesozoic era

graben (GRA-bin) a valley formed by a downdropped fault block

granite a coarse-grained, silica-rich rock, consisting primarily of quartz and feldspars, and the principal constituent of the continents, believed to be derived from a molten state beneath the Earth's surface

granulite (GRAN-yeh-lite) a metamorphic rock comprising continental interiors

graywacke (GRAY-wah-keh) a poorly-sorted sandstone with a clay matrix

greenstone a green metamorphosed igneous rock of Archean age

groundwater water derived from the atmosphere that percolates and circulates below the surface

guyot (GEE-oh) an undersea volcano that reached the surface of the ocean, whereupon its top was flattened by erosion. Later, subsidence caused the volcano to sink below the surface

gypsum (JIP-sem) a calcium sulfate mineral formed during the evaporation of brine pools
haboob (hey-BUBE) a violent dust storm or sandstorm
half-life the time for one-half the atoms of a radioactive element to decay
halite an evaporite deposit composed of common salt
hanging valley a glaciated valley above the main glaciated valley, often forming a waterfall
helictite (HE-lik-tite) a branching calcium carbonate deposit on cave walls
hematite (HE-meh-tite) a red iron-oxide ore
hiatus (hie-AY-tes) a break in geologic time due to a period of erosion or nondeposition of sedimentary rock
horn a peak on a mountain formed by glacial erosion
horst an elongated, uplifted block of crust bounded by faults
hot spot a volcanic center with no relation to a plate boundary; an anomalous magma generation site in the mantle
hyaloclastic (hi-AH-leh-KLAS-tic) basalt lava erupted beneath a glacier
hydrocarbon a molecule consisting of carbon chains with attached hydrogen atoms
hydrologic cycle the flow of water from the ocean to the land and back to the sea
hydrology the study of water flow over the Earth
hydrothermal relating to the movement of hot water through the crust; also a mineral ore deposit emplaced by hot groundwater
hypocenter the point of origin of earthquakes; also called focus
Iapetus Sea (EYE-ap-I-tus) a former sea that occupied a similar area as the present Atlantic Ocean prior to Pangaea
ice age a period of time when large areas of the Earth were covered by massive glaciers
iceberg a portion of a glacier calved off upon entering the sea
ice cap a polar cover of snow and ice
igneous rocks all rocks solidified from a molten state
ignimbrite (IG-nem-brite) a hard rock composed of consolidated pyroclastic material
impact the point on the surface upon which a celestial object has landed, creating a crater
inselberg (IN-sell-berg) isolated upland standing above the general level of the surrounding plains
interglacial a warming period between glacial periods
intertidal zone the shore area between low and high tides
intrusive any igneous body that has solidified in place below the Earth's surface
iridium (I-RI-dee-em) a rare isotope of platinum, relatively abundant on meteorites

island arc volcanoes landward of a subduction zone, volcanoes parallel to a trench, and above the melting zone of a subducting plate

isostasy (eye-SOS-teh-see) a geologic principle that states that the Earth's crust is buoyant and rises and sinks depending on its density

isotope (I-seh-tope) a particular atom of an element that has the same number of electrons and protons as the other atoms of the element, but a different number of neutrons; i.e., the atomic numbers are the same, but the atomic weights differ

jointing the production of parallel fractures in rock formations

kame a steep-sided mound of moraine deposited at the margin of a melting glacier

karst a terrain comprised of numerous sinkholes in limestone

kettle a depression in the ground caused by a buried block of glacial ice

kimberlite (KIM-ber-lite) a volcanic rock composed mostly of peridotite, originating deep within the mantle, which brings diamonds to the surface

Kirkwood gaps bands in the asteroid belt that are mostly empty of asteroids due to Jupiter's gravitational attraction

knoll (nole) a small, rounded hill

laccolith (LA-keh-lith) a dome-shaped intrusive magma body that arches the overlying sediments and sometimes forms mountains

lacustrine (leh-KES-trene) inhabiting or produced in lakes

lahar (LAH-har) a mudflow of volcanic material on the flanks of a volcano

lamellae (leh-ME-lee) striations on the surface of crystals caused by a sudden release of high pressures such as those created by large meteorite impacts

landform a surface feature of the Earth

landslide a rapid downhill movement of earth materials triggered by earthquakes and severe weather

lapilli (leh-PI-lie) small, solid pyroclastic fragments

lateral moraine the material deposited by a glacier along its sides

Laurasia (lure-AY-zha) a northern supercontinent of Paleozoic time consisting of North America, Europe, and Asia

Laurentia (lure-IN-tia) an ancient North American continent

lava molten magma that flows out onto the surface

limestone a sedimentary rock consisting mostly of calcite from shells of marine invertebrates

liquefaction (li-kwe-FAK-shen) the loss of support of sediments that liquefy during an earthquake

lithosphere (LI-the-sfir) the rocky outer layer of the mantle that includes the terrestrial and oceanic crusts. The lithosphere circulates between the Earth's surface and mantle by convection currents

lithospheric a segment of the lithosphere involved in the plate interaction of other plates in tectonic activity

loess (LOW-es) a thick deposit of airborne dust

magma a molten rock material generated within the Earth and the constituent of igneous rocks

megaplume a large volume of mineral-rich warm water above an oceanic rift

magnetic field reversal a reversal of the north-south polarity of the magnetic poles

magnetite a dark, iron rich, strongly magnetic mineral, sometimes called a lodestone

magnetometer a device used to measure the intensity and direction of the magnetic field

magnitude scale a scale for rating earthquake energy

mantle the part of a planet below the crust and above the core, composed of dense rocks that might be in convective flow

maria (MAR-ee-eh) dark plains on the lunar surface caused by massive basalt floods

mass wasting the downslope movement of rock under the direct influence of gravity

megalithic monuments large stones arranged for various purposes, including cultural monuments

mesa an isolated, relatively flat-topped elevated feature larger than a butte and smaller than a plateau

Mesozoic (MEH-zeh-zoe-ik) literally the period of middle life, referring to a period between 250 and 65 million years ago

metamorphism (me-te-MORE-fi-zem) recrystallization of previous igneous, metamorphic, or sedimentary rocks created under conditions of intense temperatures and pressures without melting

meteor a small celestial body that becomes visible as a streak of light when entering the Earth's atmosphere

meteorite a metallic or stony celestial body that enters the Earth's atmosphere and impacts on the surface

meteoroid a meteor in orbit around the sun with no relation to the phenomena it produces when entering the Earth's atmosphere

Mid-Atlantic Ridge the seafloor spreading ridge that marks the extensional edge of the North and South American plates to the west and the Eurasian and African plates to the east

midocean ridge a submarine ridge along a divergent plate boundary where a new ocean floor is created by the upwelling of mantle material

mima mounds piles of sediment caused by earthquakes

monadnock (mah-NAD-nock) an isolated mountain or hill rising above a lowland

moraine (mah-RANE) a ridge of erosional debris deposited by the melting margin of a glacier

moulin (MUE-lin) a cylindrical shaft extending down into a glacier produced by meltwater

mountain roots the deeper crustal layers under mountains

mudflow the flowage of sediment-laden water

nonconformity an unconformity in which sedimentary deposits overlie crystalline rocks

normal fault a gravity fault in which one block of crust slides down another block of crust along a steeply tilted plane thrust onto continents by plate collisions

nuée ardente (NU-ee ARE-dent) a volcanic pyroclastic eruption of hot ash and gas

oolite (OH-eh-lite) small rounded grains in limestone

Oort Cloud the collection of comets that surround the Sun about a light-year away

ophiolite (OH-fi-ah-lite) masses of oceanic crust thrust onto the continents by plate collisions

orogen (ORE-ah-gin) an eroded root of ancient mountain range

orogeny (oh-RAH-ja-nee) a process of mountain building by tectonic activity

outgassing the loss of gas from within a planet as opposed to degassing, the loss of gas from meteorites

overthrust a thrust fault in which one segment of crust overrides another segment for a great distance

oxbow lake a cutoff section of a river meander that forms a lake

pahoehoe lava (pah-HOE-ay-hoe-ay) a lava that forms ropelike structures when cooled

paleomagnetism the study of the earth's magnetic field, including the position and polarity of the poles in the past

paleontology (PAY-lee-on-TAH-logy) the study of ancient life forms, based on the fossil record of plants and animals

Paleozoic (PAY-lee-eh-ZOE-ik) the period of ancient time between 570 and 250 million years ago

Pangaea (pan-GEE-a) an ancient supercontinent that included all the lands of the Earth

Panthalassa (PAN-the-lass-a) a great global ocean that surrounded Pangaea

pediment an inclined erosional surface that slopes away from a mountain front and often covered with alluvium

pegmatite a granite with extremely large quartz and feldspar crystals

peneplain a land surface of slight relief shaped by erosion

peridotite (pah-RI-deh-tite) the most common rock type in the mantle

periglacial referring to geologic processes at work adjacent to a glacier

permafrost permanently frozen ground in the Arctic regions

permeability the ability to transfer fluid through cracks, pores, and interconnected spaces within a rock

pillow lava lava extruded on the ocean floor giving rise to tabular shapes

placer (PLAY-ser) a deposit of rocks left behind by a melting glacier; any ore deposit that is enriched by stream action

planetoid a small body, generally no larger the moon, in orbit around the sun. A disintegration of several such bodies might have been responsible for the asteroid belt between Mars and Jupiter

plateau (pla-TOW) an extensive region with a relatively level surface rising abruptly above the adjacent land

plate tectonics the theory that accounts for the major features of the Earth's surface in terms of the interaction of lithospheric plates

playa (PLY-ah) a flat, dry, barren plain at the bottom of a desert basin

pluton (PLUE-ton) an underground body of igneous rock younger than the rocks that surround it and formed where molten rock oozes into a space between older rocks

pothole deep depressions in the bedrock of a fast-flowing stream or beneath a waterfall

pumice volcanic ejecta that is full of gas cavities and extremely light in weight

pyroclastic (PIE-row-KLAS-tik) the fragmental ejecta released explosively from a volcanic vent

quartzite a metamorphosed sandstone

radiolarian (RAY-dee-oh-LAR-eh-en) a microorganism with a shell of silica that comprises a large component of siliceous sediments

radiometric dating the age determination of an object by chemical analysis of stable versus unstable radioactive elements

recessional moraine a glacial moraine deposited by a retreating moraine glacier

redbed red-colored sedimentary rocks indicative of a terrestrial deposition

reef the biological community that lives at the edge of an island or continent. The shells from dead organisms form a limestone deposit

regression a fall in sea level, exposing continental shelves to erosion

resurgent caldera a large caldera that experiences renewed volcanic activity that domes up the caldera floor

rhyolite (RYE-oh-lite) a volcanic rock that is highly viscous in the molten state and usually ejected explosively as pyroclastics

rhythmite (RITH-mite) regularly banded deposits formed by cyclic sedimentation

rift valley the center of an extensional spreading, where continental or oceanic plate separation occurs

rille (ril) a trench formed by a collapsed lava tunnel

riverine (RI-vah-rene) relating to a river

roche moutonnée (ROSH mue-tin-aye) a knobby glaciated bedrock surface

saltation the movement of sand grains by wind or water

salt dome an upwelling plug of salt that arches surface sediments and often serves as an oil trap

sand boil an artesian-like fountain of sediment-laden water produced by the liquefaction process during an earthquake

sandstone a sedimentary rock consisting of cemented sand grains

scarp a steep slope formed by earth movements

schist (shist) a finely-layered metamorphic rock that tends to split readily into thin flakes

seafloor spreading a theory that the ocean floor is created by the separation of lithospheric plates along midocean ridges, with new oceanic crust formed from mantle material that rises from the mantle to fill the rift

seamount a submarine volcano

sedimentary rock a rock composed of fragments cemented together

seiche (seech) a wave oscillation on the surface of a lake or landlocked sea

seismic (SIZE-mik) pertaining to earthquake energy or other violent ground vibrations

seismic sea wave an ocean wave generated by an undersea earthquake or volcano; also called tsunami

seismometer a detector of earthquake waves

shield areas of exposed Precambrian nucleus of a continent

shield volcano a broad, low-lying volcanic cone built up by lava flows of low viscosity

sill an intrusive magma body parallel to planes of weakness in the overlying rock

sinkhole a large pit formed by the collapse of surface materials undercut by the dissolution of subterranean limestone

solifluction (SOE-leh-flek-shen) the failure of earth materials in tundra

spalling the separation of successive thin layers from the bare surface of a rock

spherules (SFIR-ules) small, spherical, glassy grains found on certain types of meteorites, lunar soils, and at large meteorite impact sites on Earth

stalactite (stah-LAK-tite) a conical calcite deposit hanging from a cave ceiling

stalagmite (stah-LAG-mite) a conical calcite deposit growing from a cave floor

stishovite (STIS-hoe-vite) a quartz mineral produced by extremely high pressures such as those generated by a large meteorite impact

strata layered rock formations; also called beds

strato volcano an intermediate volcano characterized by a stratified structure from alternating emissions of lava and fragments

strewn field a usually large area where tektites are found arising from a large meteorite impact

striae (STRY-aye) scratches on bedrock made by rocks embedded in a moving glacier

stromatolite (STRO-mat-eh-lite) sedimentary formation in the shape of cushions or columns produced by slime-secreting blue-green algae (cyanobacteria)

subduction zone a region where an oceanic plate dives below a continental plate into the mantle. Ocean trenches are the surface expression of a subduction zone

subsidence the compaction of sediments due to the removal of fluids

surge glacier a continental glacier that heads toward the sea at a high rate of advance
syncline (SIN-kline) a fold in which the beds slope inward toward a common axis
taiga (TIE-gah) an extensive pine forest adjacent to the tundra
talus cone a steep-sided pile of rock fragments at the foot of a cliff
tarn a small lake formed in a cirque
tectonics (tek-TAH-niks) the history of the Earth's larger features (rock formations and plates) and the forces and movements that produce them
tektites (TEK-tites) small, glassy minerals created from the melting of surface rocks by an impact of a large meteorite
tephra (TEH-fra) all clastic material from dust particles to large chunks, expelled from volcanoes during eruptions
terrace a narrow, level plain with a steep front bordering a river
terrane (teh-RAIN) a unique crustal segment attached to a landmass
Tethys Sea (TEH-this) the hypothetical mid-latitude region of the oceans separating the northern and southern continents of Laurasia and Gondwana
till sedimentary material deposited by a glacier
tillite a sedimentary deposit composed of glacial till
transform fault a fracture in the Earth's crust along which lateral movement occurs. They are common features of the mid-ocean ridges
transgression a rise in sea level that causes flooding of the shallow edges of continental margins
trapps (traps) a series of massive lava flows that resembles a staircase
trench a depression on the ocean floor caused by plate subduction
tsunami (sue-NAH-me) a seismic sea wave produced by an undersea or near-shore earthquake or volcanic eruption
tuff a rock formed of pyroclastic fragments
tundra permanently frozen ground at high latitudes and elevations
unconformity an erosional surface separating younger rock strata from older rocks
uniformitarianism the idea that the slow processes that shape the Earth's surface have acted essentially unchanged throughout geologic time
upwelling the process of rising as in magma or an ocean current
varves thinly laminated lake bed sediments deposited by glacial meltwater
ventifact (VEN-te-fakt) a stone shaped by the action of windblown sand
volatile a substance in magma such as water and carbon dioxide that controls the type of volcanic eruption
volcanic ash fine pyroclastic material injected into the atmosphere by an erupting volcano
volcanic bomb a solidified blob of molten rock ejected from a volcano
volcano a fissure or vent in the crust through which molten rock rises to the surface to form a mountain

BIBLIOGRAPHY

THE EARTH'S CRUST

Allegre, Claud J. and Stephen H. Snider. "The Evolution of the Earth." *Scientific American* 271 (October 1994): 66–75.

Burchfiel, B. Clark. "The Continental Crust." *Scientific American* 249 (September 1983): 130–142.

Dickinson, William R. "Making Composite Continents." *Nature* 364 (July 22, 1993): 284–285.

Fisher, Arthur. "Alaska Down Under?" *Popular Science* 228 (June 1988): 10–12.

Francheteau, Jean. "The Oceanic Crust." *Scientific American* 249 (September 1983): 114–129.

Hoffman, Paul F. "Oldest Terrestrial Landscape." *Nature* 375 (June 15, 1995): 537–538.

Howell, David G. "Terranes." *Scientific American* 253 (November 1985): 116–125.

Jones, David L., et al. "The Growth of Western North America." *Scientific American* 247 (November 1982): 70–84.

Kerr, Richard A. "Coming up Short in Crustal Quest." *Science* 254 (December 6, 1991): 1456–1457.

Kunzig, Robert. "Birth of a Nation." *Discover* 11 (February 1990): 26–27.

Preiser, Rachel. "The First Land." *Discover* 16 (December 1995): 32–33.

Shurkin, Joel and Tom Yulsman. "Assembling Asia." *Earth* 4 (June 1995): 52–59.

Taylor, S. Ross and Scott M. McLennan. "The Evolution of Continental Crust." *Scientific American* 274 (January 1996): 76–81.

Weiss, Peter. "Land Before Time." *Earth* 8 (February 1998): 29–33.

Zimmer, Carl. "Ancient Continent Opens Window on the Early Earth." *Science* 286 (December 17, 1999): 2254–2256.

EROSION AND SEDIMENTATION

Abelson, Philip H. "Climate and Water." *Science* 260 (January 27, 1989): 461.

Ambroggi, Robert P. "Water." *Scientific American* 243 (September 1980): 101–115.

Berner, Robert A. and Antonio C. Lasaga. "Modeling the Geochemical Carbon Cycle." *Scientific American* 260 (March 1989): 74–81.

Broecker, Wallace S. "The Ocean" *Scientific American* 249 (September 1983): 146–160.

Gibbons, Boyd. "Do We Treat Our Soil Like Dirt?" *National Geographic* 166 (September 1984): 353–388.

Glanz, James. "Erosion Study Finds High Price for Forgotten Menace." *Science* 267 (February 24, 1995): 1088.

Green, D. H., S. M. Eggins, and G. Yaxley. "The Other Carbon Cycle." *Nature* 365 (September 16, 1993): 210–211.

Monastersky, Richard. "The Case of the Missing Carbon Dioxide." *Science News* 155 (June 12, 1999): 383.

Pinter, Nicholas and Mark T. Brandon. "How Erosion Builds Mountains." *Scientific American* 276 (April 1997): 74–79.

Talbot, Christopher J. and Martin P.A. Jackson. "Salt Tectonics." *Scientific American* 257 (August 1987): 70–79.

TYPE SECTIONS

Badish, Lawrence. "The-Age-of-the-Earth Debate." *Scientific American* 261 (August 1989): 90–96.

Barnes-Svarney, Patricia. "In Search of Ancient Shores." *Earth Science* 40 (Spring 1987): 22.

Goetz, Alexander F.H. "Geologic Remote Sensing." *Science* 211 (February 20, 1981): 781–790.

Hallet, Bernard and Jaakko Putkonen. "Surface Dating of Dynamic Landforms: Young Boulders on Aging Moraines." *Science* 265 (August 12, 1994): 937–940.

Kerr, Richard A. "Geology Near, Far, and Long Ago." *Science* 286 (November 12, 1999): 1279–1281.

Maslowski, Andy. "Eyes on the Earth." *Astronomy* 14 (August 1986): 9–10.

Miller, Martin. "Missing Time." *Earth* 4 (October 1995): 58–60.

Monastersky, Richard. "Ancient Ocean Upheaval Marks the Spot." *Science News* 136 (July 22, 1989): 61.

Simon, Cheryl. "In With the Older." *Science News* 123 (May 7, 1983): 300–301.

———. "The Great Earth Debate." *Science News* 121 (March 13, 1982): 178–179.

Wood, Dennis. "The Power of Maps." *Scientific American* 268 (May 1993): 89–93.

FOLDING AND FAULTING

Bird, Peter. "Formation of the Rocky Mountains, Western United States: A Continuum Computer Model." *Science* 239 (March 25, 1988): 1501–1507.

Frohlich, Cliff. "Deep Earthquakes." *Scientific American* 260 (January 1989): 48–55.

Gonzalez, Frank I. "Tsunami!" *Scientific American* 280 (May 1999): 56–65.

Gore, Rick. "Our Restless Planet." *National Geographic* 168 (August 1985): 142–179.

Green, Harry W. II. "Solving the Paradox of Deep Earthquakes." *Scientific American* 271 (September 1994): 64–71.

Harrison, T. Mark, et al. "Raising Tibet." *Science* 255 (March 27, 1992): 1663–1670.

Johnston, Arch C. and Lisa R. Kanter. "Earthquakes in Stable Continental Crust." *Scientific American* 262 (March 1990): 68–75.

Kerr, Richard A. "Delving into Faults and Earthquake Behavior." *Science* 235 (January 9, 1987): 165–166.

Molnar Peter. "The Structure of Mountain Ranges." *Scientific American* 255 (July 1986): 70–79.

Monastersky, Richard. "Mountains Frozen in Time." *Science News* 148 (December 23 & 30, 1995): 431.

Murphy, J. Brendan and R. Damian Nance. "Mountain Belts and the Supercontinent Cycle." *Scientific American* 266 (April 1992): 84–91.

Ruddiman, William F. and John E. Kutzbach. "Plateau Uplift and Climate Change." *Scientific American* 264 (March 1991): 66–74.

Stein, Ross S. and Robert S. Yeats. "Hidden Earthquakes." *Scientific American* 260 (June 1989): 48–57.

IGNEOUS ACTIVITY

Berreby, David. "Barry Versus the Volcano." *Discover* 12 (June 1991): 61–67.

Coffin, Millard F. and Olav Eldholm. "Large Igneous Provinces." *Scientific American* 269 (October 1993): 42–49.

Fischman, Josh. "In Search of the Elusive Megaplume." *Discover* 20 (March 1999): 108–115.

Francis, Peter and Stephen Self. "Collapsing Volcanoes." *Scientific American* 256 (June 1987): 91–97.

Hon, Ken and John Pallister. "Wrestling with Restless Calderas and Fighting Floods of Lava." *Nature* 376 (August 17, 1995): 554–555.

Krajick, Kevin. "To Hell and Back." *Discover* 20 (July 1999): 76–82.

Rampino, Michael R. and Richard B. Stothers. "Flood Basalt Volcanism During the Past 250 Million Years." *Science* 241 (August 5, 1988): 663–667.

Rona, Peter A. "Mineral Deposits from Sea-Floor Hot Springs." *Scientific American* 254 (January 1986): 84–92.

Stager, Curt. "Africa's Great Rift." *National Geographic* 177 (May 1990): 10–41.

Vink, Gregory E., et al. "The Earth's Hot Spots." *Scientific American* 252 (April 1985): 50–57.

White, Robert S. and Dan P. McKenzie. "Volcanism at Rifts." *Scientific American* 261 (July 1989): 62–71.

Wickelgren, Ingrid. "Simmering Planet." *Discover* 11 (July 1990): 73–75.

CANYONS, VALLEYS, AND BASINS

Arce, Gary. "From Badwater to Bliss." *Earth* 4 (February 1995): 25–33.

Birnbraum, Stephen. "The Grand Canyon." *Good Housekeeping* (November 1990): 150–152.

Cann, Joe and Cherry Walker. "Breaking New Ground on the Ocean Floor." *New Scientist* 139 (October 30, 1993): 24–29.

Dietz, Robert S. and Mitchell Woodhouse. "Mediterranean Theory May Be All Wet." *Geotimes* 33 (May 1988): 4.

Hopkins, Ralph L. "Land Torn Apart." *Earth* 7 (February 1997): 37–41.

Hsu, Kenneth J. "When the Black Sea Was Drained." *Scientific American* 251 (July 1984): 53–63.

Jordan, Thomas H. and J. Bernard Minster. "Measuring Crustal Deformation in the American West." *Scientific American* 259 (August 1988): 48–55.

Macdonald, Kenneth C. and Paul J. Fox. "The Mid-Ocean Ridge." *Scientific American* 262 (June 1990): 72–79.

Monastersky, Richard. "What's New in the Ol' Grand?" *Science News* 132 (December 19 & 26, 1987): 392–395.

Pratson, Lincoln F. and William F. Haxby. "Panoramas of the Seafloor." *Scientific American* 276 (June 1997): 82–87.

Parfit, Michael. "Timeless Valleys of the Antarctic Desert." *National Geographic* 194 (October 1998): 120–135.

Prestrong, Ray. "It's about Time." *Earth Science* 42 (Summer 1989): 14–15.

Radok, Uwe. "The Antarctic Ice." *Scientific American* 253 (August 1985): 98–106.

Stager, Curt. "Africa's Great Rift." *National Geographic* 177 (May 1990): 10–41.

Steinhorn, Ilana and Joel R. Gat. "The Dead Sea." *Scientific American* 256 (October 1983): 102–109.

Weisburd, Stefi. "Sea-Surface Shape by Satellite." *Science* 129 (January 18, 1986): 37.

DESERTS AND SEACOASTS

Bower, Bruce. "Shuttle Radar Is Key to Sahara's Secrets." *Science News* 125 (April 21, 1984): 244.

Folger, Tim. "Waves of Destruction." *Discover* 15 (May 1994): 68–73.

Fritz, Sandy. "The Living Reef." *Popular Science* 246 (May 1995): 48–51.

Hecht, Jeff. "Death Valley Rocks Skate on Thin Ice." *New Scientist* 248 (September 20, 1995): 19.

Idso, Sherwood B. "Dust Storms." *Scientific American* 235 (October 1976): 108–114.

Mack, Walter N. and Elizabeth A. Leistikow. "Sands of the World." *Scientific American* 275 (August 1996): 62–67.

Nori, Franco, et al. "Booming Sand." *Scientific American* 277 (September 1997): 84–89.

Norris, Robert M. "Sea Cliff Erosion: A Major Dilemma." *Geotimes* 35 (November 1990): 16–17.

Pendick, Daniel. "Waves of Destruction." *Earth* 6 (February 1997): 27–29.

Pennisi, Elizabeth. "Dancing Dust." *Science News* 142 (October 3, 1992): 218–220.

Raloff, Janet. "Holding on to the Earth." *Science News* 144 (October 1993): 280–281.

Stevens, William K. "Threat of Encroaching Deserts May Be More Myth Than Fact." *The New York Times* (January 18, 1994): C1 & C10.

Szelc, Gary. "Where Ancient Seas Meet Ancient Sand." *Earth* 5 (December 1997): 78–81.

Zimmer, Carl. "How to Make a Desert." *Discover* 16 (February 1995): 51–56.

GLACIAL TERRAIN

Allen, Joseph Baneth and Tom Waters. "The Great Northern Ice Sheet." *Earth* 4 (February 1995): 12–13.

Alley, Richard B. and Michael L. Bender. "Greenland Ice Cores Frozen in Time." *Scientific American* 279 (February 1998): 80–85.

Broeker, Wallace S. "What Drives Glacial Cycles." *Scientific American* 262 (January 1990): 49–56.

Clark, Peter U. "Fast Glacier Flow Over Soft Beds." *Science* 267 (January 6, 1995): 43–44.

Hallet, Bernard and Jaakko Putkonen. "Surface Dating of Dynamic Landforms: Young Boulders on Aging Moraines." *Science* 265 (August 12, 1994): 937–940.

Hoffman, Paul F. and David P. Schrag. "Snowball Earth." *Scientific American* 282 (January 2000): 68–75.

Kerr, Richard A. "Iceland's Fires Trap the Heart of the Planet." *Science* 284 (May 14, 1999): 1095–1096.

Kimber, Robert. "A Glacier's Gift." *Audubon* 95 (May–June 1993): 52–53.

Krantz, William B., Kevin J. Gleason, and Nelson Cain. "Patterned Ground." *Scientific American* 259 (December 1988): 68–76.

Mathews, Samual W. "Ice on the World." *National Geographic* 171 (January 1987): 84–103.

Mollenhauer, Erik and George Bartunek. "Glacier on the Move." *Earth Science* 41 (Spring 1988): 21–24.

Monastersky, Richard. "Stones Crush Standard Ice History." *Science News* 145 (January 1, 1994): 4.

———. "Hills Point to Catastrophic Ice Age Floods." *Science News* 136 (September 30, 1989): 213.

Moran, Joseph M., Ronald D. Stieglitz, and Donn P. Quigley. "Glacial Geology." *Earth Science* 41 (Winter 1988): 16–18.

Waters, Tom. "A Glacier was Here." *Earth* 4 (February 1995): 58–60.

Weisburd, Stefi. "Halos of Stone." *Science News* 127 (January 19, 1985): 42–44.

CAVES AND CAVERNS

Bahn, Paul G. "Ice Age Drawings on Open Rock Faces in the Pyrenees." *Nature* 313 (February 14, 1985): 530–531.

Begley, Sharon and Louise Lief. "The Way We Were." *Newsweek* (November 10, 1986): 62–72.

Bolton, David W. "Underground Frontiers." *Earth Science* 40 (Summer 1987): 16–17.

Bower, Bruce. "Cave Evidence Chews up Cannibalism Claims." *Science News* 139 (June 1, 1991): 341.

Gorman, Christine. "Subterranean Secrets." *Time* (November 30, 1992): 64–67.

Hansen, Michael C. "Ohio Natural Bridges." *Earth Science* 41 (Winter 1988): 10–12.

Lipske, Mike. "Wonder Holes." *International Wildlife* 20 (February 1990): 47–51.

Monastersky, Richard. "Volcanoes Under Ice: Recipe for a Flood." *Science News* 150 (November 23, 1996): 327.

Naeye, Robert. "The Strangest Volcano." *Discover* 15 (January 1994): 38.

Oliwenstein, Lori. "Lava and Ice." *Discover* 13 (October 1992): 18.

Palmer, Arthur N. "Paleokarst Yields Diagenetic Clues." *Geotimes* 40 (September 1995): 9.

Petrini, Cathy. "Heart of the Mountain." *Earth Science* 41 (Summer 1988): 14–18.

Young, Patrick. "Life Without Light." *Earth* (December 1996): 14–16.

COLLAPSED STRUCTURES

Decker, Robert and Barbara Decker. "The Eruptions of Mount St. Helens." *Scientific American* 244 (March 1981): 68–80.

"Facing Geologic and Hydrologic Hazards." *US Geological Survey Professional Paper 1240–B.* Government Printing Office, 1981.

Friedman, Gerald M. "Slides and Slumps." *Earth Science* 41 (Fall 1988): 21–23.

Holzer, T.L., T.L. Youd, and T.C. Hanks. "Dynamics of Liquefaction during the 1987 Superstition Hills, California Earthquake." *Science* 144 (April 7, 1989): 56–59.

Kerr, Richard A. "Volcanoes With Bad Hearts Are Tumbling Down All Over." *Science* 264 (April 29, 1994): 660.

Marsden, Sullivan S., Jr. and Stanley N. Davis. "Geological Subsidence." *Scientific American* 216 (June 1967): 93–100.

Monastersky, Richard. "When Mountains Fall." *Science News* 142 (August 29, 1992): 136–138.

———. "Soil May Signal Imminent Landslide." *Science News* 134 (November 12, 1988): 318.

Peterson, Ivars. "Digging into Sand." *Science News* 136 (July 15, 1989): 40–42.

Shaefer, Stephen J. and Stanley N. Williams. "Landslide Hazards." *Geotimes* 36 (May 1991): 20–22.

Weisburd, Stefi. "Sensing the Voids Underground." *Science News* 130 (November 22, 1986): 329.

Zimmer, Carl. "Landslide Victory." *Discover* 12 (February 1991): 66–69.

METEORITE IMPACT CRATERS

Alvarez, Walter and Frank Asaro. "An Extraterrestrial Impact." *Scientific American* 263 (October 1990): 78–84.
Binzel, Richard P., M. Antonietta, and Marcello Fulchignoni. "The Origins of the Asteroids." *Scientific American* 265 (October 1991): 88–94.
Desonie, Dana. "The Threat from Space." *Earth* 5 (August 1996): 25–31.
Gehrels, Tom. "Collisions with Comets and Asteroids." *Scientific American* 274 (March 1996): 54–59.
Grieve, Richard A. F. "Impact Cratering on the Earth." *Scientific American* 262 (April 1990): 66–73.
Hildebrans, Alan R. and William V. Boynton. "Cretaceous Ground Zero." *Natural History* 104 (June 1991): 47–52.
Kerr, Richard A. "Testing an Ancient Impact's Punch." *Science* 263 (March 11, 1994): 1371–1372.
Kerr, Richard A. "New Source Proposed for Most Common Meteorites." *Science* 273 (September 6, 1996): 1337.
Monastersky, Richard. "Target Earth." *Science News* 153 (May 16, 1998): 312–314.
———. "Shots from Outer Space." *Science News* 147 (January 28, 1995): 58–59.
Morrison, David. "Target Earth: It Will Happen." *Sky & Telescope* 79 (March 1990): 261–265.
O'Keffe, John A. "The Tektite Problem." *Scientific American* 239 (August 1978): 116–125.
Perth, Nigel Henbest. "Meteorite Bonanza in Australian Desert." *New Scientist* 129 (April 20, 1991): 20.
Roach, Mary. "Meteorite Hunters. *Discover* 18 (May 1997): 71–75.
Sharpton, Virgil L. "Glasses Sharpen Impact Views." *Geotimes* 33 (June 1988): 10–11.
Stone, Richard. "The Last Great Impact on Earth." *Discover* 17 (September 1996): 60–71.

UNUSUAL FORMATIONS

Aydin, Atilla. "Evolution of Polygonal Fracture Patterns in Lava Flows." *Science* 239 (January 29, 1988): 471–475.
Broad, William J. "Life Springs Up in Ocean's Volcanic Vents, Deep Divers Find." *The New York Times* (October 19, 1993): C4.
Cook, Patrick. "The Movable Earth Puzzle." *Science 86* 7 (May 1986): 80.
Daniel, Glyn. "Megalithic Monuments." *Scientific American* 243 (July 1989): 78–90.

Dvorak, John J., Carl Johnson, and Robert I. Tilling. "Dynamics of Kilauea Volcano." *Scientific American* 267 (August 1992): 46–53.

Fryer, Patricia. "Mud Volcanoes of the Marianas." *Scientific American* 266 (February 1992): 46–52.

Folger, Tim. "The Biggest Flood." *Discover* 15 (January 1994): 37–38.

Goodwin, Bruce K. "The Hole Truth." *Earth Science* 41 (Summer 1988): 23–25.

Hekinian, Roger. "Undersea Volcanoes." *Scientific American* 251 (July 1984): 46–55.

Lewis, G. Brad. "Island of Fire." *Earth* 4 (October 1995): 32–33.

Monastersky, Richard. "The Light at the Bottom of the Ocean." *Science News* 150 (September 7, 1996): 156–157.

Peck, Dallas L., Thomas L. Wright, and Robert W. Decker. "The Lava Lakes of Kilauea." *Scientific American* 241 (October 1979): 114–128.

Stimac, Jim. "The Strangest Place on Earth." *Earth* 6 (April 1998): 72–75.

INDEX

Boldface page numbers indicate extensive treatment of a topic. *Italic* page numbers indicate illustrations or captions. Page numbers followed by *m* indicate maps; *t* indicate tables; *g* indicate glossary.

D

E

F

V

W

X

Y

Z